The Human Being, the World and God

Anne L.C. Runehov

The Human Being, the World and God

Studies at the Interface of Philosophy of Religion, Philosophy of Mind and Neuroscience

Springer

Anne L.C. Runehov
Department of Systematic Theology
Uppsala University
Uppsala, Sweden

ISBN 978-3-319-44390-4 ISBN 978-3-319-44392-8 (eBook)
DOI 10.1007/978-3-319-44392-8

Library of Congress Control Number: 2016952417

Printed on acid-free paper

This Springer imprint is published by Springer Nature
The registered company is Springer International Publishing AG Switzerland

Dedicated with love to my dear husband Hans for his endless love and encouraging support.

Preface

Nine years has passed since my very first book, *Sacred or Neural? The Potential of Neuroscience to Explain Religious Experience*, was published. This was, it is needless to say, a major moment in my life. In 2013, the *Encyclopedia of Science and Religions* was finally published, after 7 years of hard work, not least because this enterprise was done alongside other publications, in which I act as author, editor, and both. After the publication of the encyclopedia, I realized that the time had come to write a monograph again. Hence, I started to gather ideas and research I had been performing since my dissertation in 2004.

The Human Being, the World and God mirrors the research I did as a postdoctoral fellow at the Copenhagen University during 2006–2010. The first 2 years were funded by the Uppsala University, the last 2 years by the Copenhagen Research Program of Excellence *Naturalism and Christian Semantics*, led by Niels Henrik Gregersen and Troels Engberg-Pedersen. The project I was working on at that time was called "Empathy in the Age of Neuroscience."

While empathy is a major subject matter of this manuscript, it also embodies some of my contributions as a coordinator together with Hans-Ferdinand Angel (coordinator in chief and founder of the project), Rüdiger Seitz, and Peter Holzer, to the research project *The Structures of Creditions*. This research project, which started 2010, is funded by the Karl-Franzens University of Graz. The fifth anniversary was celebrated with, amongst other things, the first edition *Processes of Believing: The Acquisition, Maintenance, and Change in Creditions* (Eds. Angel, H-F, L., Oviedo, R., Paloutzian, A.L.C., Runehov, and R. Seitz, Springer, 2016).

However, besides these major projects, the book also reflects many years of philosophical inquiries of which some have been published, some have been presented at conferences and seminars, and some which have simply remained in a notebook.

Since the beginning of my doctoral studies, I have been working within the field of science and theology/religion. Since then, my main topics of focus have been neuroscience, philosophy of religion, and philosophy of mind. Studying empathy led me to reflect more and more about what it is to be a human being. There are, however, so many aspects of human being; it felt like I was entering through the door of human life itself. The question, what it is to be human, became too big. Hence, I decided to have the book divided into three main parts, revolving around three main questions.

The first question is what is human being? The second one is what is it to be a human being? The last question became can it, in the age of science, still be argued that there is something special or specific about being human? To try to answer these questions, I had to consider not only neuroscience but also computer science and quantum physics. In other words, I had to return to previous philosophical times, picking up ideas and research I performed before I entered the debate of neuroscience and religion. This became an exciting journey, which I hope will reveal the importance of several academic disciplines being perceived as having vital contributions to make to one another. Of course, there are many questions left open, many problems still to be solved, but my hope is that my contribution may be inspiring for further studies, communication, and research.

Uppsala, Sweden Anne L.C. Runehov

Acknowledgments

I am privileged and very grateful to a host of people who have helped me in various ways in the course of the writing of this work.

I owe my gratitude to Niels Henrik Gregersen and Troels Engberg-Pedersen for their intellectual and inspiring way of engaging in my postdoctoral project on Empathy in the Age of Neuroscience. In regard to this, I also want to thank the members of the seminar Naturalism and Christian Semantics for their constructive discussions of part of my work, Tilde Bak Halvgaard, Gitte Buch-Hansen, Stefan Frederik Mortensen, Nordgaard Svendsen, René Rosfort, Lars Sandbeck, Johanne Stubbe T. Kristensen, and Runar Thorsteinson.

I also owe special gratitude to Hans Ferdinand Angel, Lluis Oviedo, Ray Paloutzian, and Rüdiger Seitz for reading and discussing part of the book, but also for encouraging me to continue my research. Not being a neuroscientist myself, I am especially thankful to Carol Albright, Jean Decety, Thilo Hinterberger, Andrew Newberg, Rüdiger Seitz, Ranganatha Sitaram, Michael Spezio, and Harald Walach for commenting, discussing, and reading these particular parts of my work and for recommending me adequate research done in the field.

Others colleagues I could discuss my work with on various occasions during all those years and to whom I want to pay my gratitude are John Albright, Thomas Anderberg†, Jennifer Baldwin, Justin Barrett, Jan Olof Bengtsson, Pat Bennett, Carl Reinhold Bråkenhielm, John Hedley Brooke, Jørgen Bo Christensen, Philip Clayton, Ron Cole-Turner, Willem Drees, Celia Deane-Drummond, Dirk Evers, Olof Franck, Michael Fuller, Antoon Geels, Peter Nicolai Halvorsen, Edward Harris, Jan Olav Henriksen, Eberhard Herrmann, Antje Jackelén, Roland Karo, Chris Knight, Lotta Knutsson-Bråkenhielm, Alex Kohav, Anne Kull, Javier Leach, David Lorimer, Hubert Meisinger, Nancey Murphy, Kees van Kooten Niekerk, Ted Peters, Gregory Peterson, Robert John Russel, Knut-Willy Sæther, Stefan Schmidt, LeRon Shults, Taede Smedes, Christopher Southgate, Neil Spurway, Mikael Stenmark, Catharina Stenqvist†, Charles Taliaferro, John Teske, Christine Tind Johannessen-Henry, Roger Trigg, and Wentzel Van Huyssteen.

I am also very grateful to Joyce Brooks and Andrew George Mattocks-Lewis, for taking care of the English language. It is needless to say that I want to say thanks to my dear family and friends especially my husband Hans Runehov, my son Thierry Galle, and my brother Clement Matthys for always supporting and encouraging me.

Contents

Introduction

What Is a Human Being?

Heb. 2:6–8 asks:

> What is mankind that you [God] are mindful of them, a son of man that you care for him? You made them a little lower than the angels; you crowned them with glory and honor and put everything under their feet.

According to the Holy Bible then, human beings are precious to God in a way that God wants to take care of them. God did not give them the same status as angels but nevertheless put them on top of creation. According to some understandings of the Bible, this implies that human beings have a particular responsibility, to care for God's creation. Human beings became God's employees, co-workers, and cocreators. However, this means that human beings need special capacities, for example, the capacity to analyze, reflect, make decisions, and possess awareness in relation to their own person and others. In order to express such capacities, Joel Green (2008) refers to Rudolf Bultmann's (1884–1976) dictum. According to Bultmann:

> Man does not have a soma, he is soma. Indeed, man, his person as a whole, can be denoted by soma. Man is called soma in respect to his being able to make himself the object of his own action or experience himself as the subject to whom something happens. He can be called soma, that is, as having a relationship to himself – a being able in a certain sense to distinguish himself from himself. The human person does not consist of two (or three) parts, then, but is a living whole. (2008: Chap. 1).

The Holy Quran also puts humans above other creations, but, where the Christian God made humans a little bit lower than angels, Allah made humans superior to them. What the Bible and the Quran have in common is that beings are given all attributes necessary in order to be able to take care of creation as a whole:

> He taught him [Adam also meaning human beings] the attributes of things and their descriptions and their characteristics, [this] signifies the vast capacity of man and the superiority of his knowledge to that of the angels. Or, the reference may be to the faculty of speech, which is the real source of the excellence of man above the whole of creation. (Maulana 2011, Section 4, 31b.).

Buddhism seems to have another opinion on what it is to be a human being. According to Buddhism:

> [a] spirit becomes embodied so as to constitute a human being. The spirit of Buddha's Law [...] is at work as a power enabling a human being to be a human being in the genuine sense. In other words, if we suppose that the spirit is the forming principle of a human being, then the spirit of Buddha's Law makes its appearance by becoming a human being. (Nishitani and Carter 2006: 121).

Hence, within Buddhism, there is no creation as we know it through the Christian or Muslim Holy scriptures. Human beings are descended spirits, and their main task is to return to the spiritual world in, at least, a more mature spiritual manner and finally, by way of different incarnations, to return as a full Buddha spirit. The common term is to return as enlightened.

If we take a look into the world of philosophy, we once again get different views of what it is to be a human being. The question "what it is to be a human being?" has followed the history of philosophy since its very beginnings. In some way we know what we are, but how do we explain ourselves? We experience that we have a body that moves in space and time and that we have a mind that also moves in space and time. Sometimes we may experience ourselves moving beyond both space and time, e.g., when we are dreaming or when we are aware that we are dreaming and are able to direct our dream at will (lucid dreaming).

Another example of experiencing ourselves beyond space and time is when we are dwelling in a state of higher consciousness. Many such experiences have been documented by people who practice meditation. Typically, people report that they do not experience space or time. Neither do they experience body sense and content. Instead they experience complete unity, or as some report, they experience pure consciousness. Frederick Travis and Graig Pearson have studied such reports by people who, by way of Transcendental Meditation, experienced "consciousness itself," i.e., self-awareness isolated from the processes and object of experience. They write:

> Pure Consciousness is "pure" in the sense that it is free from the processes and contents of knowing. It is a state of "consciousness" in that the knower is conscious through the experience, and can, afterwards, describe it. The "content" of pure consciousness is self-awareness. In contrast, the content of normal waking experiences are outer objects or inner thoughts and feelings. (2000: 79)

In contrast to experiences during a higher meditative state, we experience the world around us as separate from ourselves, but we experience the world within us as a unique part of ourselves. Nevertheless, this unique inner part of ourselves allows us to understand other human beings, in a way that they "have to" be like us. This implies that our inner self does not stand alone, but is somehow connected to other inner selves. We also project our inner experiences onto (other) animals and even onto the world of plants. We talk to our pets and plants as if they were like us. We have this feeling of belonging, of being united. Perhaps this is why the pre-Socratic philosophers sought to find a first principle that would connect all there is (Gill and Pellegrin 2006).[1]

For Thales of Miletus (625–545 BC), this first principle was water. Water was the basis of all things. Anaximander of Miletus (610–546 BC) argued that the first principle cannot be defined (*apeiron*). It is eternal and has no qualities in itself but allows for primary opposites such as cold and hot. Anaximenes (585–524 BC) reasoned that the first principle had to be air. To Heraclitus of Ephesus (535–475 BC), the first principle had to be ethereal fire. His idea was that everything derives from and returns to fire. All things are in a perpetual flux structured as logos.[2] The human soul mirrors this logos. The logos structures the ever-changing processes of the universe. "You could not step twice into the same river; for other waters flow on" (Curd and Granham 2008: 173). God became the eternal unity with Xenophanes of Colophon (570–475 BC). In his view, God permeated the universe and governed it by his thought. With this view, Xenophanes became the first to endorse pantheism. Later, Anaxagoras of Clazomenae (510–428 BC) would agree upon there being an ordering principle (God, or divine Mind, *nous*), but he also claimed there to be a material substance. With him, the mental and the material principles became qualitatively distinct. The Eleatics were skeptical about the senses. In other words, senses could not be trusted in order to find the truth. The truth could only be mediated by the *nous*.

The Sophists on the other hand relied entirely on the senses, arguing that what we perceive and act upon is merely mediated by our senses (this idea has been pursued and further developed by David Hume and, e.g., Putnam; see below). People's thoughts and opinions became standard. One could perhaps say that they were the first to think in terms of, if not anti-realism, at least internal realism (e.g., Putnam).[3] Socrates (470/469–399 BC) would accept this line of thinking but questioned people's beliefs. His method of investigating was dialectic, known as *elenchus* (inquiry or cross-examination) and *maieutics* (the art of giving birth). "The only thing I know is that I do not know anything" were Socrates' famous words with which he meant, if we can accept this, we open ourselves to the truth (in Plato 1922). We could say that it is Socrates who taught philosophers to think critically about the human being, the world, and the ultimate world. It was Socrates who taught philosophers to search for soundness of reasoning. Socrates was not so much concerned with the world, the universe, or the ultimate world; rather, he was concerned with what human beings thought about it all, whether there was consensus in their way of thinking. His disciple Plato (428–348 BC) combined Socratesian thinking and method with other philosophical ideas, for example, with the Eleatic idea of the One (God) and with Heraclitus' principle of the perpetual flux. The objects of sense became imperfect copies of their forms, of which the highest is the form of the

[1] For further reading on ancient philosophy, see *A Companion to Ancient Philosophy* (eds. Marylouise Gill and Pierre Pellegrin, 2006. The Oxford Handbook of Presocratic Philosophy. Eds. Partricia Curd and Daniel W. Graham, 2008).

[2] Meaning word, account, principle, plan, formula, measure, proportion, reckoning, reason

[3] To name some Sophists: Protagoras, Gorgias, Hippias, and Prodicus

"Good," the first cause of being and knowledge. Only by reasoning, which is the soul's activity within itself, *divorced* from the senses, is it possible to obtain true being and knowledge about the Good. This line of thinking changed with Plato's student, Aristotle (384–322 BC), but it would be of tremendous influence for, not least, Saint Augustine and the whole idea of theism.

In his *Treatise on the Soul* (*Gr. peri psyche, Lat. De Anima*), Aristotle suggested three types of soul, the vegetative, the sensitive, and the rational soul. Human beings possess a rational soul which is capable of nourishing itself, like the vegetative soul, and of experiencing sensations and moving locally, like the sensitive soul, but on top of these capacities, the rational soul has the capacity to experience and compare other forms of life. To Aristotle, the soul was simply the form of a living being. Hence, the form of a human being is not different from other living forms except that it is rational as well. To the question how body and soul are related, Aristotle's answer was that this is a trivial question about which we should not spend time worrying. He compared it to the relationship between the wax and the candle and meant that whether the wax and the candle are one or not is not a big issue. It is not clear though whether Aristotle really saw body and soul as one unity, i.e., as two sides of the same coin or whether they are separable. He did not endorse substance dualism like Plato (and onward); instead, from his hylomorphism he argued that we do not think that the Hermes shape persists after the bronze is melted and recast, so we should not think that the soul survives the demise of the body. Nevertheless, Aristotle did not escape dualism all together since he also argued that it is possible that some parts of the soul may, in the end, be separable from the body, because they are not really part of the body. What he referred to is his concept of mind, intellect, or reason *(nous)* which he saw to be an exceptional capacity of the soul. It is "the part of the soul by which it knows and thinks" (Aristotle 2005). To him, the mind is what differentiates a human being from an animal. It is what makes humans human. Besides understanding, the mind is responsible for planning and deliberating, analyzing different options, and so forth. He distinguished between a practical mind (intellect or reason) and a theoretical one. Little did Plato and Aristotle know which consequences their philosophy of mind would come to have.

Indeed, jumping over Cartesian (and other forms of dualism), it is due to post-Cartesian dualism, enlightenment, and postmodernism that we have come to perceive human beings as minds on the one hand and as bodies on the other. Furthermore, due to the enlightenment and postmodernism, we have come to divide the whole human enterprise into public-primary and private-secondary domains. The private-secondary realm stands for "all that the public rejects: the affections, relationships, caring, [...]" (Hopkins 1997: 75). The public-primary realm is considered the realm of "knowledge, objectivity, science [...]" (Hopkins 1997: 75). A consequence is that the physical "body" has gained public-primary interest, while the "mind" has become, if not a part of the body (brain), then a property of the personal-subjective realm. However, to explain the nonphysical in terms of functions of the physical may solve the mind-body-dualist problem, but it is hardly a justified interpretation of what a human being is.

What then is a human being? How can we understand this complex species? I do not believe anyone can answer these questions, but that does not imply that we should give up researching the human being. Whether this book sheds new light upon the subject matter of being human and to what extent I leave to the reader to decide. It certainly is my aim to turn previous thinking upside down and investigate what thoughts are worth preserving and what new thoughts can be added in order to deepen our understanding of the human species.

Parts and Chapters

Part I on *human being* includes three chapters.

The first chapter, "A Two- and Threefold Self," is built around four questions, firstly, what happens in the brain when we behave and experience in a certain way. In other words, why neurologically do we behave and experience as we do? Secondly, what would be the relationship between a subjective self and the brain? Thirdly, who or what might the self be; what is its function? Fourthly, why is there something we refer to as a self? Also, the term *being* is analyzed. What do we mean by human being? A distinction is made between being as *ens* and being as *esse*. The self is postulated as a (strong) emergent threefold self, consisting of an objective neural self, a subjective neural self, and a subjective transcendent self. The function of the objective neural self is to neurologically sustain the subjective selves. The function of the subjective neural self then is to express the neural self. Finally, the task of the subjective transcendent self is to be the essential observing subjective self, transcending the former two. The subjective transcendent self is seen as the part of the self that always was and always is itself, irreducible to neither the neural self nor the subjective neural self. By way of mutual causation, the three elements of the self result in the emergent process of the whole self (ES).

The second chapter, "The Human Experiencer," has as its main argument that (human) beings cannot not experience. This idea is inspired by Sartre who argued that the human being cannot not choose (Sartre 1943). To experience it is argued is the sine qua non of human being (and perhaps of other species). That human beings cannot not experience does not mean that the reality of the experiences is as it is experienced. Nevertheless, all experiences are at least subjectively real. Human beings are experiencers and as such transcend different levels of reality, including Ultimate reality. They conceive different types of reality during the course of their lives. They also share their experiences and create concepts and conceptions, which makes it possible to determine what the experiences are about. It is argued that also experiences of Ultimate reality are real in the same sense as are all human experiences. However, that does not imply that the object of experience *exists* independently of the experience. In what sense experiences are real is investigated by way of neurological studies.

The third and last chapter, "Human Uniqueness," asks the question what is it to be a human being compared to other species and compared to artificial intelligences.

In other words, what are the similarities and differences between humans and other animals and humans and humanoids? The uniqueness of the human race has been discussed since the establishment of Charles Darwin's (1809–1882) theory of evolution, in 1859, announcing that the human race evolved from the same original cell as all other animals. Humans are not the "crown" of creation; there is no crown and no creation. Rather, biological principles such as randomness, adaption, and natural selection led to the evolution of different species including the human species. Questions that are raised are whether nonhuman animals are also social animals or does the human social ontology remain to be unique? What external and/or internal information processes underpin such social reality? It will be showed that, compared to nonhuman animals, humans seem to have specific features unique to the species. Amongst other things, these specific features have inspired computer scientists to develop humanlike devices. The question is now how humans are androids? The chapter on human uniqueness is meant to function as a bridge between the questions what human beings are in relation to themselves and what they are in relation to the world.

The second part on *the world* contains four chapters.

Chapter 4, "Understanding Reality," investigates different philosophical understandings of reality. Is the world out there as we perceive it? Is the world out there a construction of our mind or both? This chapter investigates different philosophical understandings of reality; it investigates the philosophical debate called realism. Since none of these concepts are seen to be complete, yet another type of realism which I call extended realism is suggested which includes three dimensions to how human beings perceive reality: (1) a measurable or observable reality, (2) a creational reality, and (3) a phenomenological reality.

Chapter 5, "Mindreading," concerns empathy and compassion. Also here different views are analyzed and evaluated. For example, studies performed by neuroscientists, cognitive scientists, psychologists, and philosophers are taken under a philosophical magnifier. Especially the Mirror Neuron Theory is examined on its validity to explain empathy. Therefore, a distinction is made between a *closed mirror neuron view* and an *open mirror neuron view.* Inspired by Alvin Goldman (2006), who divided mindreading into high-level and low-level mindreading, I found it fruitful to also divide empathy into high-level and low-level empathy.

Chapter 6 is called "Free Will, Responsibility and Moral Evil." Philosophers have made a distinction between moral and natural evil. It is argued that this distinction is not enough to explain human evil in the world. Hence, a further distinction is made. Hence, four types of moral evil will be suggested: (1) pure accidental moral evil, (2) belief-based accidental moral evil, (3) active but not entirely responsible moral evil, and (4) active and responsible moral evil.

Also, what could be the neurological underpinnings are studied as well as psychological traits. It is argued that human moral acts are determined by their culture and psychological and/or neurological condition. However, this does not mean that people cannot be held responsible for their moral actions. Rather, it means that there seems to be a scale of responsibility corresponding to which types of moral evil are performed.

Chapter 7, "Human Time," asks the question whether time as we understand it actually exists and how can we understand the flow of time? A distinction is made between studied time and experienced time. Different explanations of time are presented. Finally, we will arrive at an understanding of time that fits with the human apprehension of it. Again, this chapter can be seen as the bridge between Part II, on the world, and Part III, God, which concerns human being's relationship to God or Ultimate reality.

Part III only contains one chapter, "God-Human-God Relationship," that leads us through different types of atheism as well as different models of God or Ultimate reality in order to evaluate how God or Ultimate reality could act in the world. Both possible divine action and divine time are considered.

This philosophical endeavor ends with a final conclusion and suggestions for further research.

The Human Being, the World and God is a philosophical analysis of what it is to be a human being in all of its aspects. It is my hope that this piece of research will once again reveal the importance of interdisciplinary research having vital contributions to make to one another and as such that it will contribute to the debate of science and religion. I do not have any ambitious aims; rather, I hope that my work can be seen as yet another piece of the big puzzle called *human being*. The method used is, once again, a conceptual and argumentative philosophical one as it is typically used in analytical philosophy. As material, I used parts of previous published articles, non-published ideas, lectures and writings, and, it is needless to say, additional literature in terms of books, anthologies, and articles reflecting the different subjects concerned with what it is to be a human being.

Part I
Human Being

Look into your heart and feel
Look into your soul and see
Look into you mind and listen
Whatever you feel, whatever you see, whatever you hear
Is part of being human

Anne L.C. Runehov

Introduction

The first part of this manuscript will concentrate on human being in relation to itself. This part contains three specific chapters, 'A Two- and Three-fold Self', 'The Human Experiencer', and 'Human Uniqueness'.

In the first chapter, four questions function as guidelines for tackling the philosophical problem of what we understand as the self. The first question concerns what happens in the brain when we behave and experience in a certain way. In other words, why neurologically do we behave and experience as we do? The second question concerns the relationship between a subjective self and the brain. The third question searches to answer, who or what the self might be, what is its function? The last question is related to the purpose of the self, i.e. why there is something we refer to as a self? A model of a three-fold self will be suggested, which implies that (1) there is only one closed emergent self, which includes both the brain and the mental processes, in which through the mutual causations of hierarchies, new distinct mental and neural properties emerge. (2) The mental activity supervenes on the neural and corresponds to different correlating altered neural activity. This concerns both specific and complex pathways between the mental and the physical. (3) The mental properties are real in the same sense as the non-mental are, i.e. they follow the same temporal and special order. Also, the term *being* is analyzed. What do we mean by human being? A distinction is made between being as *ens* and being as *esse.*

The second chapter, 'The Human Experiencer' has as its main argument that (human) beings cannot not experience. This idea is inspired by Sartre who argued that the human being cannot not choose. To experience it is argued is the sine qua non of human being (and perhaps of other species). That human beings cannot not experience does not mean that the reality of the experiences are as it is experienced. Nevertheless, all experiences are at least subjectively real. Human beings are experiencers and as such transcend different levels of reality, including ultimate reality. They conceive different types of reality during the course of their lives. They also share their experiences and create concepts and conceptions, which makes it possible to determine what the experiences are about. It is argued that also experiences of ultimate reality are real in the same sense as are all human experiences. However, that does not imply that the object of experience *exists* independently of the experience. In what sense experiences are real is investigated by way of neurological studies.

The third and last chapter, 'Human Uniqueness' asks the question what it is to be a human being compared to other species and compared to artificial intelligences. In other words, what are the similarities and differences between humans and other animals, and humans and androids? The uniqueness of the human race has been discussed since the establishment of Darwin's theory of evolution, in 1859 announcing that the human race evolved from the same original cell as all other animals. Humans are not the 'crown' of creation; there is no crown and no creation. Rather, biological principles such as randomness, adaption and natural selection have led to the evolution of different species including the human species. Questions that are raised are whether non-human animals are also social animals or does the human social ontology remain unique? What external and/or internal information processes underpin such social reality? It will be shown that humans seem to have specific features unique to the species compared to non-human animals. Amongst other things, these specific features have inspired computer scientists to develop human-like devices. The question is now, how human are androids?

The chapter on human uniqueness is meant to function as a bridge between the questions what human a being is in relation to itself and what it is in relation to the world.

Chapter 1
A Two- and Threefold Self

Introduction

Before beginning this chapter, one term has to be explained, namely, 'being'. What is meant by being? Indeed, the English term being is confusing, which has led to the problem that it is difficult to know what exactly is meant by being, especially when it comes to scientific research. Clarifying a term is important when one's research crosses the border of other disciplines or when one tries to make scientific finding understandable for non-scientists. Terms in need of clarification are, amongst others, experience, mind, consciousness, awareness, self, intelligence, affection and emotion, what is called the ego, the self and being.

The Basic Problem

Due to post-Cartesian dualism,[1] the enlightenment and post-modernism we have come to perceive ourselves as minds on the one hand and as bodies on the other. Actually, we have come to divide the whole human enterprise into private-secondary and public-primary domains. The private-secondary realm stands for "all that the public rejects: […] affections, relationships, caring, […]" (Hopkins 1997: 75). The public-primary realm is considered the realm of "knowledge, objectivity, science […]" (Hopkins 1997: 75).

One unfortunate consequence is that the physical 'body' has gained public-primary interest while the 'mind' has become, if not a part of the brain, then a property of the personal-subjective realm. However, to explain the non-physical in terms of functions of the physical may solve the mind-body-dualist problem, but is hardly

[1] I mean after-Cartesian dualism because, Cartesian dualism is, according to me misunderstood. Furthermore it is not my aim to dig into dualism more than necessary for my purpose.

A.L.C. Runehov, *The Human Being, the World and God*,
DOI 10.1007/978-3-319-44392-8_1

a justified interpretation of what human being is. For instance, if mind or consciousness derives solely from physical properties of the brain, exactly from which properties of the brain does it then derive? If, on the one hand, mind or consciousness is located in the part of the brain that is considered to be active during some specific mental experience, should we then search for mental states in the part of the brain associated with mental experience? If, on the other hand, mind and consciousness are regarded as being spread all over the brain, should we then search for these mental states all over the brain? Hence, we need to agree whether brain, mind, consciousness and awareness are one and the same thing, totally different things or two sides of the same coin called *being*.

I vote for the latter option, mind and brain are two sides of the same coin. I also argue that there is mutual causation between the physical and the mental that influence each part in a real sense. Furthermore, that there is some kind of overlap between the physical and the mental. Imagine spinning a coin vertically and you will notice that the two sides become entangled.

Being and to Be

Trying to define human *being,* I suggest that we regard *being* in two different ways: as *to be (Lat. Esse)* and as *a being (Lat. Ens).*

A being in its meaning of ens then comprises everything that is not purely mental, i.e. the neurology, biology and physiology of that being. It tells us how that being's brain is constructed, how that being's DNA map looks, which physical and biological condition that being has. We know that the neural system (as a system) is similar in all creatures. However that does not mean that the neural system is the same in all creatures. There are important differences but these have not so much to do with the system itself, rather species have different levels of neural complexity. The brain function diverging human beings the most from other species is the advanced cerebral neocortex. This is the part of the nervous system that enables humans to create language, art, myth and culture, and to make plans and create a self-image as well as a worldview. Fish have no neocortex at all; reptiles and birds have but a poor beginning. The larger the proportion of neocortex in the brain, the more it correlates with a number of social variables. Humans have the largest neocortex. For example, chimpanzees have "30:1 ratio of neocortical gray matter to the size of the medulla in the brainstem, while humans have 60: 1" (et al. 2002). A more recent study supported these findings and let Semendeferi et al. conclude that "human evolution, after the spit from the common ancestor with chimpanzees, was accompanied by discrete modifications in local circuitry and interconnectivity of selected parts of the brain" (2011:1494).

This implies that the *Ens*-part of the human being already reveals what is specific for the human species. There are also differences in DNA between the species. While humans and chimpanzees share 98.8 % of their DNA the species are very different in several ways. This is because two identical stretches of DNA can work very differently, humans and chimpanzees use their genes in different ways, and even though the same genes are expressed in the same brain in both species, they are expressed in different amounts. The evolution of the human species simply looks different and is far more developed and polished. Thus, what makes human beings authentically human can already be shown by studying the neurological-physiological aspects of being, i.e. by studying a human being as *Ens*.

This raises the question why we should be interested in the other aspects of being, namely those involving being, esse? The answer is that being as ens does not tell us much about the human way of living, an individual's needs, or what it means to be a human being. It does not tell us much about the phenomenology of a human being. In short, it does not tell us much about a human being's being. It is here that the cultural, social, religious and personal comes in. Where ens refers to the individuals, esse refers to the individuals as they relate to themselves, the world and God or ultimate reality (Also atheists relate to ultimate reality when they argue that there is no such reality).

Human beings are not only advanced neocortices and DNA (ens), they are mothers and fathers, colleagues and friends; they love and hate, and strive for better life-conditions. They paint, write poetry, create music build monuments and advanced houses, build societies and institutions, and many are religious. In other words, they also are 'esse'. One can look at the human esse from different angles, one is from the point of view we call the self or 'I', another is from the point of view of experiences and yet another is from the point of view of uniqueness.

The Self

Explaining what is meant by a self is a thorny issue. Indeed, on the one hand humans experience having some kind of a self, some kind of an 'I' that 'feels' like a deeper or elevated part of the mind and body. On the other hand, until now nobody was able to actually discover such a self or 'I'. Indeed, there may be as many explanations of the self as there are explainers. For centuries, psychologists and philosophers tugged at their hair trying to solve the mystery of the self. Some lost their hair, others went grey but no full explanation has been given. We ended up with different ideas of how a self could be understood.

Let me just name a few as a matter of introduction. Some equate the self with the person. However, philosophically, the concept of person is problematic as well. Hence, instead of the term person, the term individual was introduced. In other words, the self means how individuals comprehend themselves. Others choose to avoid the complications surrounding the term person by comparing the self with one's identity, e.g. like a mental unique personal number. Harald Walach refers to the self as being aware of oneself and knowing that fact (2007). Again others define the self as consciousness, which is yet another very complex and difficult philosophical issue.

For instance, Edmund Husserl (1859–1938) talked about an inner awareness that is always present (1984). Awareness (Bewusstsein), according to Husserl can be seen as, firstly "der gesamte reelle phänomenologische Bestand des empirichen Ich, als Verwebung der psychischen Erlebnisse in der Einheit des Erlebnisstroms", secondly "inneres Gewahrwerden von eigenen psychischen Erlebnissen" and thirdly "zusammanfassende Bezeichnung für jederlei 'psychische Akte' oder, intentional Erlebnisse" (1984: 356). Hence, to Husserl awareness is everything having to do with awareness of one's mental experiences, its role is to gather and interweave every kind of mental act or intentional mental experiences. The mental acts are always present, or one could say real, which however does not mean that the objects of the acts really exist.

Michael Lockwood compares the self with a searchlight and argues that "[…] what we see are the objects that the searchlight illuminates for us. We do not see the searchlight. Nor do we see the light: merely what the light reveals" (1989: 169).

Some Philosophers of Mind avoid the problem of the self by simply stating that it does not exist; it is an illusion; it is a function of the brain. Neurological cases such as split-brain and schizophrenia show that there cannot be any self. (Dennett 1992). Owen Flanagan agrees with Dennett that the self is constructed but argues that it is not simply a useful fiction and offers another idea of the self: "the self as a many-leveled reality […] is constructed rather than given, in which activities at each level have some autonomy and yet are related to each other" (Flanagan 1992 in Barbour 1999: 274). Constructivists and illusionists lean on David Hume's idea that he could never catch himself. According to Hume,

> For my part, when I enter most intimately into what I call myself, I always stumble on some particular perception or other, of heat or cold, light or shade, love or hatred, pain or pleasure. I never can catch myself at any time without a perception, and never can observe anything but the perception.... If anyone, upon serious and unprejudiced reflection, thinks he has a different notion of himself, I must confess I can reason no longer with him. All I can allow him is, that he may be in the right as well as I, and that we are essentially different in this particular. He may, perhaps, perceive something simple and continued, which he calls himself; though I am certain there is no such principle in me (1978: I, IV, sec. 6).

The list of ideas on the notion of the self is almost endless. Nevertheless, in my opinion, there is more to say. I do not maintain that how I see the self is the correct one, however, I believe my notion of the self to be more illustrative than previous versions of the self.

Hence, my aim is *not* to try to explain the self once and for all, but to illustrate its function. I will do this by relying on neurological findings during clinical examinations of individuals with brain dysfunctions. A first such dysfunction that, in my opinion, illustrates the function of a self is Alzheimer's disease (AD).

People who are suffering from Alzheimer's disease suffer from progressive memory impairment; disordered cognitive function; altered behavior including paranoia, delusions and loss of social appropriateness; a progressive decline in language function and motor functions (e.g. orientation). People who have the diagnosis of Alzheimer's disease (AD) are constrained by a decaying brain. Since 2003, neuroscientists are by way of imaging technologies able to discover where exactly the β-amyloid plaque, which is implicated in the disease's progression, is formed, (Klunk 2003). This was a major step forward, but they still do not know how to cure the illness. A study performed in 2013 could show robust evidence that there might be a connection between depression later in life and the development of vascular dementia and Alzheimer's disease (Diniz et al. 2013: 329–335).

Alzheimer's disease certainly brings the matter concerning the nature and meaning of human life itself to a head. It is argued that the person slowly disappears. Before the illness, these people had, as anybody else, different character traits, such as being intelligent, caring, cultural, religious, proud, etc. Where has all that gone? What becomes of the persons and their memories? If there is nothing more to the persons than a decaying brain, how should we consider the personal relationship of the sick person to themselves, the world and God or ultimate reality if it is maintained that there is no person left? Sadly, what is often overlooked is that the person continues to have feelings all throughout the course of the illness, at all stages. The question David Keck raises in *Forgetting Whose We Are,* is, "[Can there be] a metaphysical basis for the human person that this disease does not destroy?" (1996: 38–39).

With all respect to and admiration of the advances within neuroscience to understand and finally and, hopefully find a cure for this most horrendous disease, neuroscientists are not looking to answer Kerck's metaphysical questions. What we would need is an approach that brings together the being as ens and being as esse of that person which could show how these two aspects of being interact.

The question about what it is to be a human being concerns not only the neurology, physiology, biology but also, and perhaps more importantly, how they live, understand others, themselves, God or ultimate reality; how they deal with love, joy and sorrow, pain and suffering and how they are nested within the world. We need a more holistic approach that brings together the different aspects of the person and also shows how these different aspects interact. Hence three questions need to be answered.

Firstly, *what* happens in the brain when we behave and experience in a certain way i.e., *why neurologically do we behave and experience as we do*? Secondly, *how might the subjective self be related to the brain*? Lastly, *who* or *what might the self be*?

Why Neurologically Do We Behave and Experience as We Do?

To obtain an answer to that question, clinical and experimental neuroscientific approaches to an explanation of the self are considered. It is argued that the neurology of the self is complex; the subjective self seems to be neurologically sustained by a neural hierarchy as well as by specific neuro-correlations.

To the question why we neurologically behave and experience as we do, Daniel Dennett answers, because of natural selection (Mother Nature) designed our brain and nervous system in this way, though not deliberately (1993: 171–182).

Unfortunately, such a reductive view is not very helpful in understanding what we call the self or the 'I', because it fails to consider the mutual causal affection that seems to occur between the neural and the mental. Hence, we need to redefine the question as; how more exactly does the interaction between the mental and the neural work? The self refers to our experiences, our relations with and understanding of us, the world and, for many people, God or ultimate reality, our feelings and emotions, our attitudes and behavior, our thinking, dreaming, and so forth. Therefore we expect the underlying and correlating neural activity to be complex as well, if it is to sustain the self in all its expressions.

Indeed, neuroscientists explain the self neurologically in terms of complexity, embodied within the entire hierarchical system across multiple hierarchical neural levels. I suggest such a complexity to be *autopoietic complexity* (Gregersen 2008),[2] because, roughly, such a complex system cannot be understood merely through its parts; it is autonomous, multi flexible and, in principle, non-predictable. In my opinion, it seems that the neuro-correlations involved in all possible expressions of the self as reported by the patients or participants of the experiments are complex in just such a way.

Concerning the hierarchical neural levels, neuroscientists Todd E. Feinberg and Julian Paul Keenan remark that we should not understand this hierarchy in terms of a *non-nested hierarchy*, because that would imply that lower and higher levels of hierarchy are physically independent entities. Such is not the case according to these neuroscientists (2005: 663). Therefore, they argue, for example, that Descartes' view that the neural substrate of the unified self *converges* at a particular place in the brain has to be incorrect (2005: 663). Instead, Feinberg and Keenan suggest a *nested hierarchy,* which implies that the elements comprising the lower levels of hierarchy are physically combined or nested within higher levels to create increasingly complex wholes. In other words, there is a process of emergence steered by multiple causations between the higher and lower levels of the self. Hence, instead of independence we have a mutual dependency (Feinberg and Keenan 2005: 663). This implies that low- and high-level neurons contribute to consciousness and mental unification is possible. Intentional action is embedded within the entire hierarchical system across hierarchical levels.

Not only the unification of the self, but also its subjectivity (analyzed subjectively), as well as its location, are neurologically understood as a nested neural hierarchy. However, the subjectivity of the self, when analyzed *objectively*, i.e. when neuroscientists map the different brain functions, is related to a non-nested hierarchy. In other words, lower and higher levels of hierarchy are studies as physically independent entities. (Feinberg and Keenan 2005: 664). Every complex subjective self is related to a complex nested neural hierarchy (Feinberg and Keenan 2005: 673).

Suppose you are in the process of thinking, would you not expect specific neural activity to correlate with your thinking process, even though it would be embedded in the nested neural hierarchy? Would you not expect this correlating neural activity

[2] Gregersen 2008. http://www.metanexus.net accessed 27 May 2015.

to be specific in the sense that it would show an unmistakable altered activity during a process of thinking compared to when there is no such process?

Indeed, within this hierarchical complexity of neural systems there are specific neural correlates that especially support the self in its different expressions. Neurologists agree that these specific neural correlates are the medial prefrontal cortex (MPC), which is significant for the theory of mind, meta-states of self-awareness (Johnson et al. 2002), thought to be necessary for reflecting upon (Fossati et al. 2003); and regulation of one's own emotional state (Ochsner et al. 2004).

When it comes to the emotional/reflective self, Feinberg, Keenan and Fossati emphasized the significance of the functions of the right frontal, temporal and parietal regions (Feinberg and Keenan 2005; Fossati et al. 2003). Furthermore, the prefrontal, frontal and parietal regions of the brain were considered to especially correlate with experiences during meditation (Kohls and Walach 2006; Newberg et al. 2001). Perhaps this is why the frontal region is understood as the brain's commanding post, a term coined by Elkhonon Goldberg in 2001. The right posterior parietal region has been found to be a particular key in the Theory of Mind (ToM), because this is significant for distinguishing cognitive thought and actions from first person perspective to third person (Decety and Sommerville 2003; Farrer et al. 2003; Ruby and Decety 2004).

From what is said above, I suggest thus far, that what we call the self is a phenomenon of emergence. Though there are several varieties of the principle of emergence, one could say that what have in common is that it is argued that the mental process is "not merely neural, [but] something new, a fresh creation" (Alexander 1920: 6–7). Inspired by the emergentist Samuel Alexander (1859–1938), I shall argue that there is but one process which being of a specific complexity has the quality of a self, 'I'. (1920: 5).[3] Furthermore, I shall argue for a strong emergence because, "if [...] mind [is] genuinely emergent, then [...] mental phenomena must play some sort of causal role; [it] must exercise causal power of [its] own" (Clayton 2004: 30).

The Emergent Self (ES) then, consists of an Objective Neural Self (ONS) and a Subjective Neural Self (SNS) between which there is mutual causation. It is this mutual causation that supplies the emergent self, the 'fresh creation' of the self, not to be reducible to the neural (subjective and objective) self. I will return to this in more depth below.

Simply put, if there is altered activity in the prefrontal lobes it *causes* the person to think and if the person is thinking it *causes* altered neural activity in the prefrontal lobes. Because neuroactivity and subjective activity never stand alone, these activities are embedded in a complex nested neural hierarchy and a complex nested subjective hierarchy between which there is also mutual causation. There is no dualism.

[3] To Alexander this specific complexity has the quality of consciousness.

The Relationship Between the Subjective and Neural Self

Neuroscientists thus dismiss a dualistic view on the basis of their findings. Indeed, dualism is a problematic viewpoint. Whether dualists defend strong or essential dualism e.g. John C. Eccles (1903–1997),[4] and René Descartes (1596–1650),[5] transcendent dualism e.g. Bonaventura (1221–1274),[6] or implicit dualism e.g. Augustine of Hippo (354–430),[7] they struggle with the problem of how to explain the relation between the mental (mind) and the physical (brain). There always seems to remain a gap between the physical and the mental on one or another level.

Still, it might not be wise to completely dismiss the dualist comprehension of the self. As Roger Trigg rightly points out "our unreflective understanding of ourselves is dualist", and that "[w]e know we are selves persisting through time and changes to the body with lasting responsibilities" (2009: 104–118). Similarly, Eugene d'Aquili and Andrew Newberg argue that human beings, having a normally functioning brain, apprehend reality through two vividly real categories. Firstly, the conscious self seems to have a light, changeable and ethereal quality often referred to as *mind, spirit or soul*. Secondly, the conscious self seems to have an external reality often referred to as *matter or material reality* (1999: 235–236).

However, this does not mean that a dualistic explanation is the best explanation of how to understand the relationship between the brain and the mind, at least not from an interdisciplinary point of view.

Nevertheless, I believe the subjective emotional description of a dualistic feeling of the self needs to be taken into account. Hence, if we want a holistic explanation of how the brain and the mental processes are related, covering both neuroscientific findings and subjective emotional descriptions, we need to find an explanation that also takes the dualistic understanding of the self into account but *without giving into dualism.*

This, I believe means that materialism or physicalism might not be a fruitful option in order to explain the brain-mind relationship, because advocates of materialism and physicalism maintain that the self can be entirely reduced to, is identical to or it is a derivative of the neural activity of the brain. While materialists argue that no mental states, processes and events exist over and above bodily states, processes and events, physicalists mean that there are only physical facts. Hence, physicalism implies, but is not implied by, materialism (Lockwood 1989: 20 and 128). The problem is that materialists and physicalists fail to address the phenomenology of the self. They are unable to tell what it is like to have such-and-such experiences, which is to deprive human beings of their experiential reality. For this reason, among others, materialistic and physicalist understandings of the relation between the mental and the physical are not fruitful.

---

[4] Eccles and Popper 1981.

[5] in Skirry 2016.

[6] Bonaventura 1990.

[7] Gabrielsson 1926.

That leaves us with non-materialistic and/or non-physicalist models of the relationship between the brain and the mind. Non-reductionists argue that there may be states, processes and events over and above bodily states, processes and events. Often non-reductionists rely on a combination of the Token Identity Theory, (Lockwood 1989) and the Principle of Supervenience, (Lockwood 1989; Murphy 1998). According to token identity theorists, particular brain-tokens *correlate* with particular mental-tokens. Together with the Principle of Supervenience, they can argue that there is a difference at the physical state level for every difference at the mental state level, on which the difference of the relevant mental state in some sense depends (Lockwood 1989: 21). However, it is important to note that, even though mental states depend on physical states, they are not reducible to them. Because of the irreducibility, philosopher Philip Clayton calls this type of supervenience *weak supervenience* (2004: 124–125). This means that there is a dependency *relationship between mental states and the brain*. Indeed, this is what neuroscientific research on the self reveals. However, as Clayton also argues, non-reductive physicalism (or non-reductive materialism) has its problems. Indeed, non-reductionists need to explain the 'over and above the physical or material', without falling into dualism, physicalism or materialism. Can this problem be solved?

Lockwood tried to solve the problem by arguing that mental states and events are real, because they follow the *same temporal and spatial order* as physical states and events. Therefore, and supported by Einstein's Special Relativity Theory, Lockwood assigns mental states independent locality (mental states must be in space given that they are in time). He writes,

> Let A be a physical event that causes a mental event B, which in turn causes a physical event C. If we know the time of occurrence and spatial locations of A and C, then we can at least place bounds upon the time of occurrence and spatial location of the mental event B: it must lie within the intersection of the forward light cone of A and the backward light cone of C (Lockwood 1989: 75).

Because Lockwood gives phenomenal qualities *real* existence (1989: 163), he challenges the philosophical view that only physical and material properties can be said to exist, because they can be traced in space and time. Lockwood's move is a fruitful one. The mental state has its legitimate place in space and time; it is real and is itself a multiple level hierarchical property. As such it supervenes upon the physical/material, correlates with it but is not reducible to it. What is not clear is whether this relation between the mental and physical also implies mutual causation (Fig. 1.1).

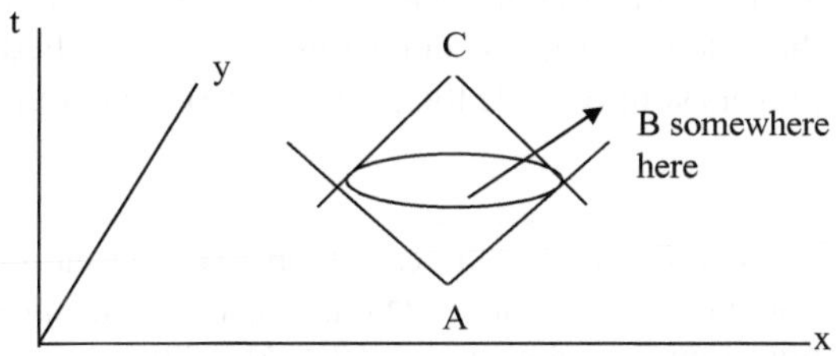

Fig. 1.1 Representation of Lockwood's idea of the spatial location of a mental event

Furthermore, Lockwood's view does not consider that and how new mental and physical properties emerge from this relation. Hence, something is still missing in this non-reductionist explanation, namely a process of emergence.

It is argued that the mind is an emergent property of the brain and, furthermore, that there is a 'mutual interdependence' between "the sustaining physio-chemical processes and the enveloping conscious qualities" (Clayton 2004: 23). As mentioned above, I rely on Alexander (1920) and argue that there is but one emergent process which being of a specific complexity has the quality of a self 'I'.[8] Furthermore, also as mentioned above, I argue for a strong emergence because, "if [...] mind [is] genuinely emergent, then [...] mental phenomena must play some sort of causal role; [it] must exercise causal power of [its] own" (Clayton 2004: 30).

The self is emergent because, "[t]he activity of the emerging human subject produces reality in an iterating, dialectical process whereby the subject is repeatedly confronted with an 'other' (*das Andere seiner selbst*) and overcomes the difference in a new synthesis that both transcends and preserves (*hebt auf*) their difference" (Clayton 2004: 79–81; Gregersen 2004: 29–31).

Not all agree with me though that mental causation is possible. Mostly, causation is understood to be directed from the material/physical to the mental; the other way around is often seen as problematic. Still, if we understand the plasticity of the brain, such opposition is difficult to understand. Indeed, our brains change continually in relation to what we experience and to experience is a mental act.

Such opposition is difficult to understand if we look at the scientific studies on piano players. Namely, scientists discovered that piano players have different brain capacity compared to people who do not play. Piano players use both their hands simultaneously while navigating 88 keys. Because of this they have to overcome something innate to almost every person, namely right- or left-handedness. But there is more to it. The notes to be played by the left hand are written in a different key than the ones to be played by the right hand. That means that not only do piano players overcome the difficulties of right- or left-handedness, they also overcome the difficulty of multitasking. The piano player *reads and plays* two different keys simultaneously. Being a piano player myself, I know, of course that managing this takes lots of dedication and training. However, that is the point; one trains how to play the instrument while at the same time training one's brain.

Commonly the depth of the brain's central sulcus is either deeper on the right or on the left side, depending on which hand is dominant. While scanning pianists, it was discovered that they had a noticeably symmetrical sulcus, even though they were born right- or left-handed. That does not imply that piano players do not have a dominant hand, most have. This proves what I said above, that the equal depth of the sulcus did not occur naturally, but resulted from the years of training. Piano players simply made their brains into better-rounded machines (Taylor Sloan 2014).[9]

---

[8] To Alexander this specific complexity has the quality of consciousness. (Alexander 1920: 5).

[9] http://mic.com/articles/91329/science-shows-how-piano-players-brains-are-actually-different-from-everybody-elses, accessed 27th May 2015.

Another important study concerning mental causation on the brain comes from Alexander Castro-Cladas et al. By way of PET imaging they studied illiterate and literate people and observed that illiterate people did not use or have access to the same brain structures that literate people have. It appeared that these structures were closed to them. Roughly the participants were twelve right-handed women from the same social and cultural background. Six illiterate women of average age 65 and six literate of average age 63. The reason for only having women in the study was because gender may influence the functional organization of language. Those who were classified as illiterate had never entered school and could not read nor write. The literate participants had 4 years of education and performed normally on reading and writing. The task was to recognize single letters or very common acronyms such as trademarks and television channels. Also pseudo words were used (i.e. words that are similar to comprehensible words but without having any meaning).

The participants were asked to repeat words or pseudo words as follows. "You are going to listen to a list of words presented one at a time that you should repeat. Some of the words are known, others you have never heard. You should repeat the words immediately and try to repeat the words exactly as you heard them" (1998: 1054–1055). All participants had to repeat words at least 90 % correctly in order to be included in the study.

The result of the repetition of comprehensible words correctly was 98 % in case of the literate women and 92 % in case of the illiterate women. This difference, even though it is rather small, reaches statistical significance. The result of the repetition of pseudo words correctly was 84 % for the literate and 33 % for the illiterate group. (Castro-Cladas et al. 1998: 1055). The conclusion drawn was that learning to read and write during childhood influences the functional organization of the adult human brain (Castro-Cladas et al. 1998: 121, 1053–1063). These findings are in agreement "with the hypothesis that experience through learning partly determines the development and organization of the human brain and, in particular, that language experience influences the functional organization of language-relevant systems" (Castro-Cladas et al. 1998: 1060).

Also meditation seems to have major causal effects on the brain efficiency and structure. Andrew Newberg and his team observed that meditators had a significantly different thalamic laterality index at baseline (before the act of meditation began) compared with the control subjects. Consequently, the differences in the thalamus led them to assume that the meditators underwent changes in their brain, because of their intense meditation practice (Newberg et al. 2001: 117–122; Runehov 2007: 155–156).

I think it is correct to suggest that mental process supervenes on the neural and corresponds with correlating altered neural activity and that mental properties are real in the same sense as the non-mental are, i.e. they follow the same temporal and special order. This implies that properties exist over and above the material and physical, but both the mental and the neural belong to one complex nested hierarchy, in which through mutual causation, new distinct mental and neural properties emerge.

I suggest we call the model of the relationship between the mind and the brain *the emergent process of the self,* which thus far contains the objective neural self and the subjective neural self between which there is mutual causation.

$$ES(ONS \leftrightarrow SNS).$$[10]

What or Who Is the Self and What Is Its Function?

In order to answer the question who or what the self might be, and what its function is, we need to look at some neurological research of the self and investigate it closer. The neuroscientific studies include both clinical and experimental studies.

Clinical Neuroscientific Studies

Clinical neuroscience has the means to investigate people with brain dysfunctions and to study what consequences these dysfunctions have for human behavior and experiences. In other words, they can investigate how important neural patterns that correlate with specific brain dysfunctions are for the human 'self'.

Frégoli and Capgras Syndromes

Frégoli and Capgras syndromes are two versions of Delusional misidentification and reduplication (DMS). Frégoli syndrome roughly implies that strangers become well known people. A nurse is seen as the patient's sister or mother etc. The Capgras syndrome is roughly the brain disorder which causes patients to misidentify people, places, objects and events. For example, doubles or impostors have replaced a familiar person. The patients cannot be convinced that the husband is who he says he is and they deny that they are mistaken when the delusion is pointed out to them. But also, the patients may misidentify a part of their own body. Shirley suffered from Asomatognosia which is related to Capgras syndrome. Hence, Capgras syndrome is about negative doubts, hypo-identification and under-personalization, while Frégoli syndrome concerns positive doubts, hyper-identification and over-personalization.

One of the patients Feinberg and Keenan examined had a large right hemisphere infarct and because of this, described a feeling of alienation from her left arm. Her name is Shirley and she describes her arm as follows (2005: 666–667).

[10] I say 'thus far' because there is still more to tell about the self.

Shirley:	*It took a vacation without telling me it was going. It didn't ask, it just went.*
Feinberg:	*What did?*
Shirley:	*My pet rock. [She lifted her lifeless left arm with her right arm to indicate what she was talking about].*
Feinberg:	*You call that your pet rock?*
Shirley:	*Yes*
Feinberg:	*Why do you call it your pet rock?*
Shirley:	*Because it does not do anything. It just sits there. [...] She belongs to me and she's a her. She's mine but I don't like her very well. She let me down. [...]*

Then Shirley held her left hand to her cheek and hugged it and kissed it and fondled it and petted it (Feinberg and Keenan 2005: 667).

Another patient, a 41-year-old man who was diagnosed with Frégoli syndrome claimed he was at work and that he was a member of the hospital staff:

Examiner:	*[...] what is your job?*
Patient:	*I am a computer person.*
Examiner:	*Where are we right now?*
Patient:	*We're at [name of his company] in New York...My office is right around the corner [pointing]. If they have problems with their computers I solve them.*
Examiner:	*[pointing to a therapist of the facility that the patient had previously claimed he knew]. And you know her?*
Patient:	*Yes [...] Her job is to do research on certain items and then bring them to [his company. [...] when she has problems with her computer she comes to me.*

Discussion

Neurologically, what can these dysfunctions tell us about the self?

As mentioned earlier, it tells us that at least one element of the self, call it the *subjective neural self,* embedded in a nested subjective hierarchy, seems to be wired to specific functions of the brain, in the cases mentioned, the right frontal, parietal and the temporal regions of the brain; these are also embedded in a nested neural hierarchy. This means that when specific neural functions become impaired the personality of the person changes in accordance with the impairment. In other words, the patients react and experience as they do because these brain regions fail to establish the appropriate ego boundaries (Feinberg and Keenan 2005: 675). Some patients suffer from a two-way ego disturbance which means that there is both an alienation from the arm (it's not my arm) as well as the projection of the arm into the environment (it belongs to my sister).

Sigmund Freud noticed that "there are cases in which parts of a person's own body, even portions of his mental life – his perceptions, thoughts, and feelings – appear alien to him and as not belonging to his ego" (1961: 13). We know that the frontal lobes control our personality; that the anterior prefrontal regions are associated with self-awareness, social interaction and judgement as well as awareness of the mental states of others, which is known as the theory of mind (ToM), (Johnson et al. 2002: 1808–1809). The ToM commonly refers to a specific cognitive capacity, namely the capacity to attribute mental states: beliefs, intents, desires, pretending, knowledge, etc., to oneself and others and to understand that others have beliefs, desires and intentions that are different from one's own.

In both Shirley's case and in the case of the 41-year old man, frontal neural activity altered in a way that caused impaired motor functions (Shirley) impaired judgement (Shirley and the 41-year old man), and impaired social behavior, interaction and judgement (the 41-year old man).

The parietal lobes have two functions, but it is the right parietal lobe that is associated with our personality. Damage to this region may cause the neglecting of parts of the body or space, cause difficulties in constructing things (constructional apraxia) and cause denial of deficits (anosagnosia). Damage to the left parietal lobe may cause difficulties with writing (agraphia), difficulties with mathematics (acalculia), difficulties with language (agraphia or aphasia) and inability to perceive objects normally (agnosia). Indeed, Shirley thinks of her arm as something alien. Right hemispheric lesions can produce significant changes in personality, which was the case for Shirley. Finally, temporal lobe activity is especially important for the formation of our personality as well as for our affective behavior. Hence, following the clinical neuroscientific observations, when something goes wrong in the neural circuits of the frontal, parietal and/or temporal regions, simultaneously, something goes wrong with a person's self-self and self-other (or self-world) experiences (Feinberg and Keenan, 2005: 673–675).[11]

Experimental Neuroscientific Studies

Face Recognition

Experimental neuroscientists, in order to understand the brain functions underlying or correlating with the subjectively experienced emotional or cognitive self-self and self-other relation, investigate healthy individuals by letting them perform (for example) emotional and/or cognitive tasks in order to understand the brain functions involved. Experimental studies, in which people were presented with both familiar faces and faces of the self, showed that the right hemisphere is twice as active when images of the self are presented compared to when images of a familiar

[11] Also the right occipital region seems to some extent to have a role to play in self-experiences.

person were shown. Similar results came from studies with familiar voices compared to the person's own voice (Feinberg and Keenan, 2005: 673–675).[12] The self to self relationship is suggested to strongly correlate with the right prefrontal and frontal regions while the right parietal regions are significant for the self-other relation (ToM) (Feinberg and Keenan, 2005: 674). Because temporal activity has to do with human emotional behavior, the right temporal lobe will show increased activity in correspondence to increased emotional behavior towards the self (compared to emotional behavior toward others). Hence, both clinical and experimental neuroscientific studies emphasize right hemispheric activity when emotional evaluation of the self is at stake.

For example, in their studies on self-reflection Johnson et al., Stuss et al. and Ochsner et al. all emphasize the significance of the medial prefrontal cortex for judgments of internal states of others guided by the own feelings of self, but argue that there is no significant distinction between the right and left hemispheres (Johnson et al. 2002).[13] However, the reason no right hemispheric activity is emphasized is because the tasks given to the participants did not necessarily trigger a subjective emotional reaction. Indeed, the task of the participants in Johnson's study was to deliver an oral statement (yes or no) to questions such as "I often forget things; I'd rather be alone; I can be relied upon; I am good at my job etc." (Johnson et al. 2002: 1809). On the other hand, when Philippe Fossati presented emotionally charged words (positive and negative) to the participants during an fMRI scan, he found, similar to Johnson and Ochsner, the medial prefrontal cortex to be significant but also emphasized the right side of the MPC, and suggested that the right MPC represents states of an emotional episodic *self* (Fossati et al. 2003: 1939–1942).[14]

The difference between Johnson's, Ochsner's and Fossati's results suggests that the *right* hemispheric regions, especially the *right* medial prefrontal cortex, are significant for the more subjective aspects, requiring a perspective, involved in emotional evaluation of the self (Ochsner et al. 2004: 1752). However, in all studies it is emphasized that the specific brain functions do not work in isolation to sustain the self. Self and other perception of emotion rely on a complex neural network. According to Ochsner the posterior cortical regions support bottom-up processing

[12]The function of the left hemisphere is associated with sequential analyses such as systematic, logical interpretation of information; interpretation and production of symbolic information language; mathematics, abstraction and reasoning and also with memory storage in a language format. The function of the right hemisphere then is associated with holistic functioning such as processing multi-sensory input simultaneously to provide a 'holistic' picture of one's environment; visual spatial skills. Holistic functions such as dancing and gymnastics are coordinated by the right hemisphere. Memory is stored in auditory, visual and spatial modalities.

[13]Hypermetabolism in the posterior cingulated gyrus was also associated with very-early Alzheimer's disease, because this is the period when patients clearly start to show difficulties with memory. This hyperbolism was also exhibited by individuals who are at genetic risk from Alzheimer's. Johnson et al. 2002: 1811. Stuss et al. 2001. Ochsner et al. 2004. Discussion of the results: pages: 1749–1750, 1752–1754, 1765.

[14]This was regardless of the positive or negative input.

(recognition of intentional behavior) while the medial prefrontal cortex supports top-down processing (attributes about mental states that guide them).

Meditation

Neurological studies done on advanced Zen-meditation also showed the right prefrontal/frontal and parietal regions to be significant (Sitaram 2013).[15] During the self-regulation of emotion and because meditation is training in focusing, altered right prefrontal, frontal and temporal activity was observed. Also Newberg found significant increased activity in the right frontal lobes (Newberg et al. 2001: 117–122). These neuroimaging studies are supported by Electroencephalogram (EEG) studies of meditation. Because decreased posterior superior parietal lobe activity is associated with decreased self-other boundaries (a mental state that typically results from the effort to silence all thought), a significant decrease in both the right and left posterior superior parietal lobes (PSPL) was observed (d'Aquili and Newberg 1999; Newberg et al. 2001: 117–122).[16] The results of the neuroimaging study is once more supported by EEG studies from which it was concluded that meditation practice increases the frequency of experiences of ego loss, implying decreased activity in the PSPL (Kohls and Walach 2006: 144).[17]

Once more, the relationship between the mental and the neural properties has been clarified; if something is affected in either of these properties, something is correspondingly affected in the other. It is now time to try to answer the third question, *who* or *what might the self be*?

The Function of the Self

In order to answer the question who or what the self might be, we need to look closer to some phenomenological explanations concerning the function of the self. In this regard what is understood as first, second and third person perspective is important.

From the analysis of clinical and experimental neuroscientific studies of the self it seems that the phenomenology of the self is vulnerable. Wittgenstein noticed this and maintained that "whatever seems to the subject right about his state of consciousness is right" may be incorrect (Wittgenstein 1953 in Guttenplan 2001: 619).

---

[15] Sitaram 2013, http://www.bme.ufl.edu/labs/wp-content/blogs.dir/12/files/2013/10/Neuroimaging Meditation.pdf Accessed 1 February 2016.

[16] For a complete analysis and evaluation of these studies, see Runehov (2007) Chap. 5 and 6.

[17] Depending on one's personal and spiritual outlook they are regarded as potentially stabilizing or destabilizing. The crucial point seems to be that those with regular spiritual practice assess experiences of ego loss much more positively.

Wittgenstein's argument is the argument against private language, of course, but nevertheless he rightly remarks that, if the private phenomenology is always correct, there would be no intra-subjectivity or intersubjectivity possible. For something to be as it is, there ought to be rules for how that something ought to be, i.e. rules that ought to be used in self-self, self-other and other-self relationships.

Remember Shirley who calls her paralyzed arm her pet rock that took a vacation without telling her. This is a clear example that one's private phenomenology can be incorrect. The self to self relationship has turned into a self to other relationship within the person's own self. Shirley still says that the pet rock belongs to her. She might still have intra-subjective rules, i.e., she might still have good reasons for why she calls her paralyzed arm her pet rock. She might compare her arm with other things she calls pet rock. However, there is no intersubjective rule that would agree with her intra-subjective rules for what concerns her arm. Concerning the 41-year-old man who suffered from the Frégoli syndrome, we might even wonder whether he has proper working intra-subjective rules when he mistakes the hospital for his office and the hospital staff for his colleagues.

How to understand this? I got some fruitful help from my husband, Hans Runehov who is a computer scientist. He argues that the brain, in normal functioning neurological circumstances seems to have some kind of *read-write authority.* In other words, there seems to be a read-write collaboration between the neural and subjective self. If we accept this view, and because of the brain dysfunction, the brains of Shirley and the 41-old man, they seem to have a *restricted write authority* or even a *read-only* authority.

Split-brain patients may perhaps best illustrate what a restricted write authority looks like. Split-brain patients' writings correlate with the hemisphere that is not associated with language. Hence, it seems that the capacity to write does not need to be associated with the capacity for phonological representation (Gazzaniga 2008). In other words, it seems that the capacity to write is independent of the neural language system. The capacity to write seems to "[…] stand alone and does not need to be part of our inherited spoken language system" (Gazzaniga 2008). Of course, due to the nested neural hierarchy correlating with the nested subjective hierarchy, one expects there to be neural support for the writing action in some way. However, it seems that this neural support does not derive from the one expected for such behavior. Hence, the mental self seems to have a restricted authority, i.e. it is able to *command* the neural circuit only in a restricted manner.

The following example of a person suffering from a dysfunctional brain illustrates a read-only contact. A reverend who was diagnosed with Alzheimer's disease, and for as long as possible wrote a diary wrote: "When the darkness and emptiness fills my mind, it is totally terrifying. I cannot think my way out of it" (Hopkins 1997: 82). It thus seems that the reverent has only a read-only authority left, which nevertheless seems to notify him of his current situation that something is definitely wrong with both the neural and subjective neural self. The question is though, what or who is that read-only authority?

It is time to ask, who or what is this *reader* seeming to observe and seeming to be aware of what is happening? Who or what is this *reader* 'who cannot *think* his way out of it'? Who or what is this *reader* realizing, concluding and crying out 'why'? Who is the *reader* aware of the darkness surrounding the mind, the *reader* that calls her arm a pet rock that does not listen to her anymore or the *reader* writing merely from the non-linguistic part of the brain? Does not that at least suggest that there is some element of the self that seems to be able to transcend both the *neural self* (the neural correlations of the self) and the *subjective neural self* (the subjective self that correlates one-to-one with the neural self)? Could one element of the self be a *subjective transcendent self* to which John H. Brooke refers as the *rational self, the suffering self, the unique self* (2006: 253–277)? Could there be a transcendent element of the self that points to another, deeper layer of a self-concept?

Let me suggest a notion of the self that is at least in accordance with the neuroscientific and philosophical analysis above. Thus far I have postulated an Emergent Self (ES) comprising an Objective Neural Self (ONS) and a Subjective Neural Self (SNS) between which there is mutual causation. However, this model seems to be in need of reconsideration. Instead of a dual-emergent self, neuroscientific results seem to point to a threefold-emergent self. The third parameter of this threefold self I suggest, I called *the Subjective Transcendent Self* (STS).

The idea of such a transcendent self is not new. Indeed, what I refer to as the subjective transcendent self has been explained in different ways. As mentioned above, Walach names it consciousness, i.e. personal awareness of being oneself and knowing that (Walach 2007: 216). Lockwood refers to it in terms of a searchlight (1989: 169). Hume could not discover impressions about himself: "I never catch myself at any time without a perception and can never observe anything but the perception" (1978: 272). Husserl speaks of an inner consciousness (inneres Bewusstsein) that is always present (1984); something he could not catch but is somehow always there. Heidegger refers to it as *Dasein (being-there)* which is the entity which in each case I myself am. Dasein is the particular being whose being is constituted by always already being-in-a-world. Dasein is always now (1993). This entity is according to Heidegger the only entity that cares about the meaning of its own being and that of others. This being is never exhausted by any conditions; it always "stands out into some other set of possibilities, so that at any given moment it is at once what it is, what it has been, and what it may yet become" (Seigel 2005: 571). "Dasein always exists beyond itself, and this way of being 'outside of itself' is its manner of being 'in and for itself', its characteristic manner of being what it is" (Seigel 2005: 571). These philosophical understandings give us a hint of what the subjective transcendent self might be, not least, the phenomenological experience of *always* being there. However, always being there does not mean that one always has full control.

Norton Nelkin distinguishes between the *being in-control* and the *being not-in-control*, which to him, is the basis for the distinction between *me* and *not-me* or *self* and *not-self*. Furthermore, according to him, there is an *underlying essential subjectivity*, an introspective capacity that makes one aware of one's distinction in one's own experience. I think that the examples above show that human beings realize that their body and their mind are out of control. To Nelkin "the subjectivity associated with a sense of control is the deepest sort of subjectivity, underlying all concept formation, including one's concept of one's very self" (Nelkin 2001: 570–575). Might this underlying essential subjectivity be understood as my suggested subjective transcendent self that, under normal circumstances has *read-write* authority (is in-optimal-control of the self and not-self); under other circumstances a *restricted read-write* authority (is in-restricted-control of the self and not-self) and under the worst circumstances merely a *read-only* authority (is not-in-control of the self and not-self but remains observant)?

The Three-Fold Self

Perhaps it is now possible to postulate a (strong) emergent threefold self, (ES) consisting of an Objective Neural Self (ONS) , a Subjective Neural Self (SNS) and a Subjective Transcendent Self (STS) . The relationship between the three elements is as follows:

$$\text{ES}\left(\text{ONS} \leftrightarrow \left(\text{SNS} \cap \text{STS}\right)\right);\ \text{STS} > \left(\text{ONS} \cap \text{SNS}\right).$$

The function of the ONS is to neurologically sustain the subjective selves. The function of the SNS then is to express the neural self. Finally, the task of the STS is to be the essential observing subjective self, transcending the former two. To put it differently, to be the self that always was and always is itself which is irreducible to either the neural self nor the subjective neural self. By way of mutual causation, the three elements of the self cause the emergent process of the whole self (ES).

However, I need to account for one possible objection, namely that there cannot be any part of the self without neural activity supporting it. I agree, but my point is that whatever neural activity may lie behind the seemingly transcending self, it is not that neural activity that neuroscientists typically associate with the subjective activity. Because the self is embedded in a nested neural hierarchy, there is somehow neural support in order to make the subjective transcendent self, able to remain observant, even during the very last stage of Alzheimer's, when almost every function of the brain has decayed. Exactly which neural function(s) are the supporting factors remains an open question. Some suggest it to be the brain stem, since this is said to be the last brain function intact before death. Another suggestion is that the underlying brain activity would be what neuroscientists have discovered and called the brain's *default mode network (DMN)*, some kind of neural dark energy (Raichle 2010: 28–33).

While a lot of questions remain open, it is argued that the brain exhibits internal activity across multiple regions while at rest. Neuroscientists suggest that this DMN prepares the brain for conscious activity and hence may provide the clue to the nature of *consciousness*. Perhaps, it may provide the clue to the nature of the STS.

> [O]ur conscious interactions with the world are just a small part of the brain's activity. What goes on below the level of awareness–the brain's dark energy, for one–is critical in provid-

> ing the context for what we experience in the small window of conscious awareness' (Raichle 2010, 33).

It seems that there is as yet no clear answer to what sustains the self (or consciousness, or awareness) until the very end of life.

Conclusion

To try to understand what it is to be a human being in all its complexity has been a philosophical endeavor for centuries. One of the queries has been to account for what is meant by what we call the self or the 'I'. I am not claiming to have tackled the problem because that is simply not possible. In this chapter I aimed at providing further distinctions and concept analyses that I hope might help research on the self, within different disciplines. I started with making a concept analysis of the English term being and suggested that we regard *being* in two different ways: as *to be (esse)* and as *a being (ens).* A being in its meaning of ens then comprises at least the neurology, and the DNA, of that being. It tells us how that being's brain is constructed, how that being's DNA map looks, which physical and biological condition that being has. However, it was argued that ens does not tell us much about the phenomenology of a human being.

Hence being understood as esse was suggested which comprises the cultural, social, religious and personal. Where a being (ens) refers to the individual, to be (esse) refers to individuals as they relate to themselves, the world and God or ultimate reality. This distinction introduced the section on the self. Where there is plenty understanding of what we call the self might or might not be, I suggested, on the basis of my reading of neuroscientific research that we could (and perhaps should) understand the self in terms of a threefold self; *the Objective Neural Self* (ONS) *the Subjective Neural Self* (SNS) and *the Subjective Transcendent Self* (STS). I suggested that the function of the ONS is to neurologically sustain the subjective selves. The function of the SNS then is to express the neural self. The task of the STS is to be the essential observing subjective self, transcending the former two. In my opinion, neuroscientific research also suggests that the complex nested hierarchical subjective self correlates with a complex nested neural hierarchy.

Hence, there is causation from the mental properties towards the physical as well as from the physical towards the mental faculties, giving rise to new and irreducible properties by way of a process of emergence.

The model of the emergent self thus implies that firstly, there is only one closed emergent self, which includes both neural and mental processes, in which through the mutual causations of hierarchies new distinct mental and neural properties emerge. Secondly, the mental activity supervenes on the neural and corresponds to different correlating altered neural activity. This concerns both specific and complex pathways between the mental and the physical. Thirdly, the mental properties are real in the same sense as the non-mental are, i.e. they follow the same temporal and special order.

Chapter 2
The Human Experiencer

Introduction

The aim of the previous chapter was to account for human being in terms of ens and in terms of esse. This definition let me dig somewhat deeper into the philosophical query of what we call the self or the 'I'. I suggeted that the self or 'I' is threefold, consisting of a neurological, a subjective and a transcendent degree of this emergent self. This chapter searches to find an answer to the question whether there might be the sine qua non of human being (in terms of esse). This philosophical question is of course not new and several answers have already been given. Nevertheless, there seems to be space for yet another enquiry on the subject matter. Hence, let us look closer to what has been said and which possibilities are still open. The title of the present chapter reveals that its main topic is about experiencing and experiences; it is about human being as experiencer. I will suggest that experiencing is the sine qua non of human esse; that a human being cannot not experience.

In Search for the sine qua non of Human Being

Several philosophers spent years pondering upon what could be special about human being (esse). I will only name a few.

Jean-Paul Sartre (1905–1980), one of the key philosophers of existentialism, reasons in *l'être et les Néant* (1943), that human beings are condemned to be free, and since there is no Creator in his view, human existence, not only precedes human essence, but is also a depressive existence. Indeed, being (esse) is to never be what one really is but to always be what one is not. Hence he seems to argue that the sine qua non of human beings is always to have to choose. Human beings possess a freedom without actually being free. Sartre argues that human beings cannot not choose.

A.L.C. Runehov, *The Human Being, the World and God*,
DOI 10.1007/978-3-319-44392-8_2

In the act of choosing, the person is alone. Their choices are absolute, unique and individual (1943).

In his *magnum opus, Die Welt als Wille und Vostellung* (1818/1819), Arthur Schopenhauer (1788–1860) argues that human being (esse) strongly correlates with the *Wille zum Leben* (will to live). However, while we commonly see a will to live as something positive and meaningful, Schopenhauer saw it as something destructive, illogical and directionless, hence, this type of willing needs to be destroyed. The will to live, he argues, is the innermost essence of human being (the sine qua non); it is without knowledge and is blind (1992). There is no distinction between bodily movement and will; there is no causality between them. The will provides a priori knowledge about the body and the body a posteriori knowledge about the will (Schopenhauer 1992: 18: 168).[1]

Schopenhauer, clearly inspired by Eastern thinking argues that this type of will is evil, and can only be terminated by way of asceticism and chastity. Only then, the will to live or the will for one-self can become the Will in Itself which is absolutely free. Roughly put, in Schopenhauer's view human beings need to depart from their being in order to be complete and free. While there is no way out of the human dilemma in Sartre's view, there is one in Schopenhauer's. Both Sartre and Schopenhauer present quite a depressive view on what human being is about.

For Sartre human being is about being free, about always having to make choices and hence never being able to be the being one is. For Schopenhauer human being is about willing to live, which implies that since he sees the world and the self as representations, human esse becomes an elusive being that misses the point of actual true life. Also Sartre's idea leads to this conclusion since human being is never what it really is.

In my opinion, neither Sartre nor Schopenhauer actually found the sine qua non of human being (esse). The reason why I think so is that human beings *can* choose not to choose (for example), and human beings *can* transcend their will to live into what Schopenhauer calls Will with a capital W. Both Sartre and Schopenhauer have provided us with a deeper insight into some of the human traits or characteristics. However, if we want to postulate a sine qua non for human being (esse) it has to be something that *cannot* change, something humans do not have the power to alter.[2]

There also are scientific explanations for what could be the most important feature of human being (esse). For example, if evolutionary biologists and advocates of evolutionary biology would suggest the sine qua non of human beings they would most probably argue that it is no different from other species in the animal kingdom. They would most probably argue that the essence of being, all being, is to survive, which amongst other things includes producing offspring. That humans have proven to be skilled survivors is because evolution provided them with an advanced neocortex, especially advanced cognitive capacities.

However, in my view, the will to survive or the drive to survive cannot be the essence, because humans, for instance, do not always choose to produce offspring.

[1] My translation from Swedish.

[2] It needs to be said that neither Sartre nor Schopenhauer uses the term sine qua non.

Some do not even want to survive. That does not mean that the drive for survival is not an important and significant trait of (human) being.

The list of examples can be made much longer, but perhaps it is enough with what is stated above and in the introduction in order to pursue my goal.

I agree with Sartre that one of the quandaries of human existence is frequently having to choose. I also agree with Schopenhauer that the will is a powerful instrument in human life. For instance, the relationship between God or ultimate reality and humans is sometimes seen as a "union of feeling and will" (Burns 2002: 15).

I agree that the neocortex and its further developments are significant for distinguishing humans from other species. It is argued that it is the human neocortex that enabled humans to talk about mind, spirit, soul, personhood, and what I postulated in the first chapter, a transcendent subjective self.

Lastly, I agree that the drive for survival is a significant feature of (human) existence. Still, in my opinion none of these suggestions could be the sine qua non of human being (esse). What these human characteristics do is serve, in one way or another, what is actually the core of human being, namely to experience.

By human experiences I intend all experiences a human being is able to have. Since a human being is related to the world, I shall designate a human experience as an experience of things and events whose existence ultimately is constituted of properties and relations, actions and interactions of whatever the physical and psychological worlds treat of (Lockwood 1989: 20).

According to Keith Yandell, a (human) experience, let us call it x, is *an aspect of the subject of experience*, let us call it y (1993: 42). Yandell maintains that "if an item A is so related to person S that *S does not exist* by itself or together with some set of truths, entails *A does not exist*, then A is an *aspect* with respect to S" (1993: 38). He argues that human beings are an aspect with respect to themselves but not to another person. Their hopes and dreams, beliefs and attitudes are aspects with respect to themselves but objects with respect to another person. It is important to note that the subject of experience is not the experiencer but is that of which an aspect relative to the experiencer is presented to him or her. Yandell continues that "[h]aving an experience is an aspect of the experiences' subject. Experiences all have subjects, though not all have objects. [That] *having an experience* is an aspect [of the experiences' subject] relative to the one who has it [...] does not entail that experiences cannot have objects" (1993: 42).

Yandell thus argues that all human experiences are aspects of a subject but do not necessarily have an *object*, let us call it y'. An illustration of a human experience that does not have any object would be experiencing thirst, because there is no object in that experience. The experience of being thirsty is an aspect of the subject of experience, namely, neural signals from the central nervous system. Experiencing drinking water on the other hand has an object, namely, water. Yandell also maintains that "*[a]n* experience is a conscious state in which a subject [a person] (phenomenologically) seems to be aware of an aspect or object" (1993: 16). By *seems* he means "how things appear, whether or not they are that way or not" (1993: 44).

For example, people who are afraid of snakes and see the tail of a cat wagging from behind some stones, thinking it is a snake, will experience their fear of snakes.

The experience of fear will be recognized as such until they find out that the snake they seemed to have experienced was actually the tail of a cat. In other words, the experiencers will know whether their experience of fear is legitimate when they receive evidence that justifies their experience of fear. I will return to the justification of experiences below.

A la Sartre, I argue, *human beings cannot not experience*. Of course all species experience, including plants. For instance, neuroscientist Thilo Hinterberger succeeded in producing music by plants. Water and electrolytes make plants grow, which causes electrical potential shifts, which can be measured by way of electrodes attached to the stem, leaves and blossoms. Hinterberger measured the electrical signal changes of various plants and flowers. The signals were then used to play various electronic instruments. Hence, each sound is initiated by the electrical signals of a plant, in other words, the music is produced by the plants (2011). It is indeed amazing to listen to plants' music, and it certainly makes one aware of the variety and wonder of life.[3]

Hence to experience is perhaps the sine qua non of all living beings. But then again, sine qua non is not equal to uniqueness. Conditio sine qua non refers to an indispensable and essential trait, without which 'something' could not be. It is in this regard that I argue that being free, always having to choose, the will to live, survival are not good candidates for this task. There may be condiciones sine quibus non in human esse. However, to experience, to be an experiencer should score highly.

The body is without doubt the sustainer or sometimes the destroyer (as for example due to brain dysfunction); and similarly, the mind contains the drives human beings need (for good and worse) in order to be. However, neither the body nor the mind by itself establishes thc essence for human esse. In my opinion then, the sine qua non of (human) existence or being (esse) is to be an experiencer. In other words, experiences are what constitute our reality. (Walach and Runehov 2010: 158). We experience all the time, from the fetus stage to dying. In doing so we transcend different levels of reality but all of them are part of our reality. With transcendent experiences I do not only intend e.g. mystical experiences or experiences of higher consciousness, but also experiencing waking up, experiencing recovery from illness, and so forth. By transcending I simply mean going from one mental state to another.

Similar to Sartre I argue that in the act of experiencing, the person is alone. Also like Sartre's choices, experiences are always absolute; they are unique and individual. Even if two people experience a headache at the same time *t*, their experience of their headache will be absolute, unique and individual. This is so because each headache requires the adaption of the point of view or perspective of the particular experiencer, i.e. the perspective provided by undergoing the headache. It is the person introspecting her or his phenomenal state of being in pain that forms the

[3]The music is produced by a sonification tool that was originally programmed for sonification of neuronal signals from the human brain. Hinterberger used it as a plant-computer musical interface.

conception of what it is like to have a headache at that particular time (Tye 1999; further discussed in Runehov 2007). And this conception of what it is like to have a headache cannot be other than absolute, unique and individual. We can understand the other's pain due to knowing that the other is 'like oneself' but we cannot share or experience the other's pain. This is why we have to account for the realness of experiencing and experiences.

Because, that esse is indispensable from experiencing does not, however, imply that experiences are real, in the sense that they convey reality as it is in itself.

The Realness of Experiences

The philosophical debate that concerns the realness of experiences has a long history.[4] On the one hand, there are those who argue that our experiences are real because they are experiences of something that exists out there (a reality independent of us). Differently put, our experiences are *true* representations of the outer reality; the object or subject of experience is exactly as we experience it.

This kind of realism is described by Hillary Putnam as, "there is exactly one true and complete description of 'the way the world is'" (1981: 49). Those defending such a view of reality advocate *naïve or metaphysical realism.* On the other hand, there are those who defend *anti-realism.* According to this view there is no reality independent of our experiences at all. Between these two opposite versions of realism we find critical -, internal, pragmatic – and non-metaphysical realism. I will return to this debate in part II when I will discuss how reality can be understood.

For now I simply wish to say something about realness of experiences as experiences. In order to do so, I want to present a fruitful distinction of the terms *concept, conception and conceiving,* a distinction suggested by Eberhard Hermann (2008).

Concept, Conception and Conceiving

The concept of water is H_2O and represents the chemical properties and necessities for water to be water. However, we do not conceive of H_2O. The conception of water then represents the ability to apply the concept water in a correct way so that it is possible to determine whether or not one is referring to water. For example, a conception of water could be that it boils at a temperature of 100 C. Conceiving water concerns how water is experienced, for example as warm, cold, wet, colorless, etc.

Colors are real from the perspective of human perception. Physicists, relying on their methodological enquiries, maintain that colors as such do not *exist* (Ward 2005: 207). What they mean is that they cannot detect colors or the sources of colors

[4] See further, Runehov 2014.

other than through human experiences, i.e. colors as conceived.[5] Because we conceive colors the way we do and because we possess the necessary cognitive capacities, we can establish its concept and conception. The concept of water is H_2O, its chemical constitution. The concept of colors is somewhat more complicated since there is no chemical nor physical form for it. What can be said is that there are physical preconditions.

> Neither objects nor light are actually 'colored' in anything like the way we experience them. Rather, color is a psychological property of our visual experiences when we look at objects and lights, not a physical property of those objects or lights. … Color is more accurately understood as the result of complex interactions between physical light in the environment and our nervous system. There may be light of different wavelengths independent of an observer, but there is no color independent of an observer, because color is a psychological phenomenon that arises only within the observer. (Palmer 1999: 97–98).

We can establish the conception of colors, namely, that colors appear once the preconditions for colors meet a living brain. We can establish that what exist are the physical and neurological preconditions for an observer to experience color.

How colors are conceived is, not really individual, but there are individual differences and of course differences between the species of the animal kingdom. Nevertheless, they are conceived as real. Perhaps the following question is a little bit Pooh philosophical but why would male birds compete with their colorful feathers if the female bird would not be able to see these grandeurs (and choose the male which to her has the most beautiful feathers)? Hence colors are indeed real *for us.*[6]

The example of colors also suggests that, in the course of our life, we "discover that there is sometimes, perhaps often, a difference between the way that the world appears to us and the way that it really is" (Ward 2005: 207). Nevertheless, the wonder of colors is not only that we experience a variety of colors, similarly to but differently from other species; we also create out of our experience of colors. Indeed, we transfer our subjective reality of colors to an objective (or at least intersubjective) reality of colors, as, for example, in arts and architecture. Hence, how we conceive of the world has an impact on reality (Walach, H., and Runehov 2010: 159). We experience water and colors in an absolute, unique and individual way, but thanks to our capacity to conceptualize, we can talk about our experiences and share them with others to some extent. This is why experiences, while being subjective in core, may become intersubjective.

Another example is the concept of a tsunami, namely, a giant wave or series of waves caused by an undersea earthquake, volcanic eruption or landslide. It is measurable but unpredictable. Its impact on land is catastrophic. However, by studying this natural phenomenon and by way of conceptualizing it carefully, we might be able to prevent catastrophes in the future. We might be able to manipulate the exposed environments in a way that minimizes the damage if a tsunami should strike again. A phenomenon such as a tsunami belongs to the type of reality which gives us both empirical and existential *resistance*, to use the term coined by

[5] This is also true for the phenomenon of gravitation and what is hypothesized as the graviton.

[6] For a more extensive discussion on the reality of colors see Byrne and Hilbert 2003.

Herrmann (2008). We cannot do other than to conceptualize it in order to understand it. The examples of water, colors and the tsunami are examples of a reality that gives us immediate resistance.

However there are measurable or observable realities that provide indirect resistance. We could call these secondary realities or realities that depend on another reality. Take for instance stemcells. Even though the term stemcells was already coined by histologist Alexander Maximow in 1908 for scientific use (see: Konstantinov 2000), it was not until the 60's that American researchers discovered how the bone marrow of mice renewed itself, and it took another 10 years before researchers discovered stem-cells in human blood (Till, J.E. and E.A. McCulloch: 1961). An important remark though is that the dimension of reality that gave us resistance was not the stemcells but illnesses such as cancer and leukemia more precisely. It was by studying this illness that stemcells were conceptualized. Today researchers have been able to map different types of stemcells and consequently, more types of illnesses can be cured with the help of these cells. Stemcells became real, with the consequence that they have impact on our reality.

To exemplify reality that gives us existential resistance and has a clear impact on our lives (our reality) is the experience of good and bad events. Those who often experience good events will most probably (but not always) project their good experiences onto their environment, while individuals who have a serious lack of such experiences are more likely (but again not always) to project their bad experiences into the world (Walach and Runehov 2010: 158). Taken to its extreme, people who have merely experienced good events in their lives, were raised under the very best and lovely circumstances, were loved and cared for as children, got the best education and only good relationships, etc., might think and believe that all people have good intentions and will not harm them. Unfortunately, that is not a realistic picture of human reality. Hence, at its extreme, those people might get into real trouble when they discover that not all human intentions are good; on the contrary, they might be wicked. On the other hand, people who have experienced a lot of cruelty and ruthlessness, who did not experience love and compassion, or being taken care of and supported, may project these experiences onto all other people, honestly believing that all people are bad people. Hence, at its extreme, they will miss the good intentions of people, miss the good fortune they so longed for but do not dare to hope for.

Extreme, yes, but there are unfortunately too many examples of such extreme cases, not least of the latter.[7] A much less extreme example is the example of nightmares. Most of us have had at least one nightmare experience and remember how scared we were, how real the experience was, how our body started to sweat because, even though we are only having a dream-experience, the experience triggers a reaction from our nervous system in the same way as if we were awake. We try to find our way out of this scary situation, implying that our neural system is fully alert. We toss and turn in our bed but experience running away or fighting. We might cry or become angry – not in a fictive sense but factually. If we wake up at that very

[7] See further below on the problem of evil.

moment, we realize that we have actually been crying when we notice that our cheeks or pillow is wet, or we discover that we are still shedding tears. It is said that our dreams are reactions to what we have consciously or unconsciously *experienced* earlier.

Hence, previous experiences trigger new ones at another level of reality, thereby triggering adequate reactions in our brain and mind.

Whatever behavior comes from whatever experience, it will bring about new experiences for us and others, shaping new behaviors shaping new experiences. However, the reality of the experiences, so it seems, is not (always) as we conceive of it.

Human beings and other animals also cannot exist without experiences. One difference between human beings and animals is that humans have the ability to 'analyze' their experiences, which, as a matter of fact, is nothing but yet another experience at another level of reality, a level of reality perhaps not (yet) achievable for other animals. Human experiencers want to understand why they experience.

They want to know whether others experience as they do. They compare experiences. Human beings experience that they are experiencing, and so on. Obviously, this raises ontological and epistemological questions.

Ontological questions require an answer as to whether our experiences are merely subjectively real or intersubjectively real. Epistemological questions concern how we can know that our experiences are reliable and which conclusions we ought to draw from them. These are questions only raised by the human species.[8]

Because of the increasing possibilities for humans to experience (environmental, cultural, personal), and the projection of these onto the natural world, the human brain developed and continues to develop accordingly, something that is necessary in order to sustain the increasing experiencing. As I mentioned in the first chapter, there is a mutual causation between brain and mind activity. Experiences entail the development of the neocortex, which led to the development of the theory of mind.[9] This capacity, it is said, is crucial for intersubjective communication (amongst other things).[10] Humans as non-humans are *experiencers*. What we call life is a chain of experiences and it has consequences for oneself and the world around us.

However, people do not only experience themselves and the world around them; they may also experience a reality that seems to transcend everyday reality. Since there are many testimonies by people who have experienced what is called divine or ultimate reality, it is only fair that these kinds of experiences are taken into account as well.

---

[8] I will return to this subject matter in the section on justifying experiences below.

[9] I explained the ToM in chapter one, but a short reminder does perhaps not hurt. The ToM is the capacity to attribute beliefs, intents, desires, pretending, knowledge, etc., to oneself and others and to understand that others may have beliefs, desires and intentions that are different from one's own. We attribute mental states to ourselves and others by way of analogical inference, i.e. the idea that other human beings are 'very like me'. x observes y, y is like x, hence x understands y.

[10] Intersubjective communication is not merely established between humans.

Experiences of Ultimate Reality

In part three which is on the on the relationship between God or ultimate reality, the world and human beings, I will present different models of God or ultimate reality. For the time being, it suffices to say that God can be understood as a personal God, distinct from the world. God can be defined in terms of the absolute ground of everything that is (and ever may be), as the creator of everything that is, as equal to the world and as both distinct from and equal to the world.

There is indeed plenty of evidence given of experiences of divine or ultimate reality. One question is how to distinguish an experience of divine or ultimate reality from an experience of everyday reality?

In *Sacred or Neural? The Potential of Neuroscience to Explain Religious Experience,* (2007) I referred to William Alston's definition of experiences of divine or ultimate reality. Following his definition once again, experiences of divine or ultimate reality are those experiences in which it seems that the divine or ultimate appears to a person in a certain way. Hence, Alston places the burden on something (taken by the experiencer to be divine or ultimate reality) presenting itself to a person's experience. Doing so has the advantage of being able to compare experiences of the divine or ultimate reality with other experiences taken by the person to be an experience of something specific (Alston 1991).[11] The difference then between experiences of everyday and ultimate reality lies in the topic of the experience. Nevertheless both kinds of experiences are real to the experiencer. One important difference is though, that very often (if not always) experiences of divine or ultimate reality have life changing consequences.

Of course, every-day experiences may also turn one's life completely upside down. When I met my husband in Brussels 1986, my life changed. Not only did I experience being in love, but this experience also made me leave my home country and move to a country of which I knew little. However, how immense this experience was, how great the love, it did not change, that much, the person I am. When it comes to experiences of divine or ultimate reality it is argued that experiencing divine or ultimate love exceeds human love by far; it changes the person sometimes in completely unexpected ways. Let us look at a few examples taken from Evelyn Underhill's 600 pages book on mysticism (1912).

It is said that Saint Francis of Assisi had difficulties whether to live the life of the world or the life of the spirit. One day when he was walking in the country side, he passed a little forgotten church (the church of S. Damiano). Struggling with his thoughts, he went in to pray. He kneeled before the Crucifix. Suddenly and unexpectedly he experienced being "smitten by unwonted visitations [and] found himself another man than he who had gone in" (Underhill 1912: 218). Taken away from all everyday reality, he experienced the picture of Christ speaking to him, calling his name: "Francis" it said, "Go, repair my House, the which as thou seest is falling into

[11] For a comprehensive study on human experiences (every-day experiences, religious experiences and psychical experiences) as well as how to explain experiences, see Runehov 2007: chap. 2.

decay" (Underhill 1912: 218). Francis could do no other than to obey. His entire worldview had changed forever; there was no more struggle as to which type of life to choose. There were no more hesitations and no more uncertainties.

Another example of such fascinating experience is the conversion of Catherine of Genoa. It is said that Catherine possessed a religious nature already from childhood but suffered a life of loneliness and was often depressed, not least due to an unhappy marriage. Being unhappily married was not exceptional at that time, especially amongst the wealthy people because wealthy parents would choose wealthy partners for their children, without consulting the child involved. Catherine was a member of a noble family, and hence could not do anything but accept her parent's choice of husband. But Catherine's spiritual longing made it all even more difficult. When she was about twenty-six, she and her sister attended a Catholic festival. We can guess that the sisters were talking about Catharine's life whereupon her sister suggested Catherine make a confession. Catherine was not convinced a confession would change her odds, but she obeyed anyway. Similarly to the experience of Saint Francis, as she knelt before a statue of Christ, her whole world changed as she experienced the "unmeasured Love of God, with so clear a vision of her own misery and her faults, and of the goodness of God, that she almost fell upon the ground" (Underhill 1912: 219). She experienced God's infinite love. Her revelation was so powerful that she did not finish her confession but ran out of the church. Her life had indeed changed completely. From that moment, she received communion almost daily, and underwent remarkable experiences. These made her to take care of the sick in a hospital in Genoa. But also her husband joined her after also having been converted. From having ruined them financially, he became a Franciscan tertiary but without joining the order. Instead the couple moved into a large hospital in Genoa and dedicated themselves to works of charity. They spent most of their lives serving the sick, not least during the plague which ravaged Genoa in 1498 and 1501.

There are plenty of similar descriptions of life-changing experiences. One finds such stories across religions as well as in non-religious worldviews. The experiences seem to come spontaneously or are obtained by way of long term meditation, yoga and prayer. They are not only found in history but also occur today. Consider the following example of transforming experiences as it was told to me by one of my informants whom I called Grace:

One Sunday, when I was about 8 years old, I was sitting in the living room on my own, reading the children's Bible. It was natural for me to do so, because I am the daughter of a minister, and I liked the Bible stories very much. Suddenly, a kind of white smoke came out of the page I was reading. Being the daughter of a minister and having heard many Biblical stories, I was not afraid at all but curiously watched what happened. The white smoke turned into figures that smiled at me and left. I remember this experience as something that really happened, and have no doubt about the reality of the spiritual figures. A year later, on an occasion when I was lying in bed making shadows on the wall with my hands (forming a duck etc.), I suddenly saw that one of my hands burned. Amazingly, my hand did not hurt at all; I was not really afraid and watched what happened. The fire disappeared as quickly as it came, leaving no marks at all. Later in life, when I became an icon painter, I remembered my burning hand and believe that this was a divine sign that my hands would become important instruments for the religious task I would be given (Runehov 2007: 122–123).

Another type of life-changing experience worth accounting for are Near-death experiences.

Near-Death Experiences

While experiences of God or ultimate reality, though heavily discussed, are quite accepted as being a part of human experience, near-death experiences are not only heavily discussed, they are very often seen as completely illusive and even nonsense. Most people accept that some people believe in a divine reality. They may have a different opinion but not many argue that religious experiences are illusive or nonsense. There are of course those who say so, e.g. Michael Persinger (1987), and Richard Dawkins (2006). When it comes to near-death experiences, more are skeptical, because these experiences are often presented as proof of an afterlife, proof of human being's immortality. We cannot prove whether or not God exists or whether or not there is an afterlife, but it is still important to look at these experiences, to try to understand them. After all, they belong to the whole package of human experiences.

Near-death experiences can be regarded as experiences of the divine or ultimate in the sense that they are profoundly transcendental. The difference is that they occur on the threshold of death. Furthermore, there are no particular personal traits, states of mind, or training required in order to have a near-death experience. The only thing needed is, indeed, to be very close to death. Such experiences are neither rare nor new. They are already documented in the Tibetan and Egyptian books of the dead, and they have been discussed ever since. These books are said to be the world's most important spiritual texts from the past. Though written in completely different areas they describe a process of transcending from this to another life. The Tibetan Book of the Dead was written in the eighteenth century, the Egyptian over 4000 year ago. The former is written as a guide for the dying, the latter as a guide for the afterlife. Interesting is that the Tibetan book gives similar accounts to those people who have had a near-death experience.

Lately, there has been an increase in reports of near-death experiences. One reason is that survival rates have improved due to contemporary medical advances, with the consequence that such experiences are well documented in contemporary clinical reports as occurring in different life-threatening events. Near-death experiences are reported in cardiac arrests (clinical death); in "shock in postpartum loss of blood or in perioperative complications; septic or anaphylactic shock; electrocution; coma resulting from traumatic brain damage; intracerebral hemorrhage or cerebral infarction; attempted suicide; near-drowning or asphyxia; and apnea" (Van Lommel et al. 2001: 2039). NDEs are also reported during the terminal phase of illness and are then called "deathbed visions" (Van Lommel et al. 2001: 2039). People surviving a near-death experience most often report that the experience left them with positive emotions toward life and the afterlife.

What then is a near-death experience? Except for the fact that it occurs at the threshold of death, the phenomenology of near-death experiences is explained in (at least) two different ways. According to Eugen d'Aquili and Andrew Newberg, near-death experiences are "unfold…according to a single pattern almost as though the prospect of death serves to release a stored, common 'program' of feelings, perceptions, and experiences" (1999: 122). Others have argued that there is no common program; rather the experience is affected by the culture and/or personal background of the experiencer. Nevertheless, there are similarities, reported aspects that are more common than others.

Indeed, often people having had such an experience report peacefulness; a dark tunnel or space; out-of-body experience; the realm of light; the realm of colors; meeting others (e.g., deceased relatives but also religious personalities); a review of past life events; and entering some unearthly realm and experiencing extrasensory phenomena. Of course, not all aspects occur during each near-death experience. Furthermore, they are described differently by the individual reporting the experience. Nevertheless, according to the literature and reports, there is one common feature, namely, the experience of peacefulness at the beginning of the experience.

Another feature that almost all report is an out-of-body experience. Several survivors report having seen themselves from somewhere above their body. Those who were in a clinical environment at that time also report, some in detail, what the medical staff did and said. Many, but not all, report entering a dark tunnel or space, which they recall as emptiness, the end of the universe, through which they are drawn toward a colorful brightness. Regardless of whether they report a journey through a darkness of some kind, most report an experience of light. This light is often described as consisting of wonderful colors and sometimes involves beautiful music. It is also in the light that some (but not all) survivors report having met with others (e.g., relatives or religious personalities). What is not always reported is a review of past life events.

Several medical studies show that age and frequency seem to be associated with the experience. The frequency of near-death experiences occurrence is highest in children, namely 85% (Morse et al. 1986: 299–318), followed by those with a mean of age of 37, namely, 48% (Sabom 1998), and 43% in people with a mean age of forty-nine (Ring 1985). In general, the frequency of near-death experiences occurrence is higher among people younger than sixty. Concerning the depth of the experiences, several studies have found that women have deeper experiences than men (Van Lommel et al. 2001: 2043). To the question why this is so, I do not have an answer.

An example of a testimony of what Pim Van Lommel called a deathbed vision goes as follows. A 69 year lady had been ill for quite a time and knew that she did not have much time left. In good Catholic tradition, she had the last rites. While the priest performed the last sacrament, she was drifting deeper and deeper into a coma, her family sitting around her bed. Some prayed; others just kept watch; all cried now and again. Candles and flowers were put around her. After 2 days, the old lady woke up. Not only did she wake up, but she was very decisive. In accordance with the typical characteristics of a near-death experience, she described the dark tunnel and the light. In the light, she said, she met Jesus, who welcomed her home. But, she continued, she did not feel like going home yet, even though it was so peaceful and beautiful 'out' there. She asked Jesus to give her some more time so that she could ensure that her youngest son would be safe. At that time he was a troubled young man in his late twenties and had given her many sleepless nights. She said that Jesus told her that she would not be successful in her goal but that he granted her wish. The next thing she remembered was coming back into her body, and that was painful, she recalled. She was given approximately a year before she was called home again. Sadly, she never managed to help her youngest son onto the safe shore (as Jesus had warned her). This lady's near-death experience shares similarities with many other near-death experiences, including the dark tunnel, the light and meeting religious personalities or spiritual beings. What she did not tell was whether she had an out-of-body experience or met deceased relatives. As with others who have had a near-death experience, this woman's story also confirms that for those who undergo these experiences, they do keep clear and sharp cognitive abilities during the experience. What is more blurred is their notion of time and space.

Melvin Morse and his colleagues studied near-death experiences of children. They interviewed eleven patients between the ages of three and six who had survived cardiac arrest and profound comas. Seven of the children remembered the following: six reported out-of-body experiences; five reported entering a darkness of which four remembered a dark tunnel; finally, three remembered deciding that they wanted to return to their body (Morse et al. 1986).

An amazing story comes from a young American woman, given the pseudonym Pam Reynolds, who underwent extremely difficult brain surgery in order to remove a brain aneurysm. The procedure included cardiac arrest, lowering her body temperature to 60 ° F, stopping her breathing, and draining the blood from her brain in order to cool it and then reintroduce it. At the point where her body temperature reached 60 ° F., all brain activity had stopped, and the aneurysm was removed. About two hours after awakening, and after the tubes had been removed from her trachea, she reported having had a near-death experience. Her experience included an out-of-body experience and moving through a dark tunnel toward a light that she described as growing larger and larger. She reported hearing her deceased grandmother's voice, encountering figures in a bright light, among whom there were deceased relatives amongst whom an uncle who convinced her to return to her body (Sabom 1998: 42–46). Reynolds proved to have had a very deep near-death experience with subsequently verified observations during the period of the flat EEG. Again, her experience is similar to others in several ways (moving from darkness to brightness, meeting deceased relatives, and out-of-body experiences).

In a prospective and longitudinal survey on near-death experiences, (Van Lommel et al. 2001), 344 patients were studied (mean age 62.2 years). All patients had been clinically dead, which Van Lommel and his colleagues authenticated mainly by electrocardiogram records. One particular patient—let us call him Sylvester—was brought into the coronary care unit by ambulance during Van Lommel's night watch.

He received artificial respiration with intubation, while heart massage and defibrillation were also applied. In the process of intubation, Van Lommel removed his upper dentures and put them on the 'crash cart'. It took about an hour and a half before the patient had sufficient heart rhythm and blood pressure to be transported to the intensive care unit. After more than a week, Van Lommel met his patient, who was back in the coronary care unit. When he saw his doctor, Sylvester told him exactly what had happened during his deep coma, including the fact that Van Lommel had removed his teeth and put them on the crash cart. He had seen himself lying in bed; he described the room correctly and in detail, as well as the people present, including Van Lommel. He also told the doctor that he had been very afraid that they would turn off the CPR and that he desperately and unsuccessfully tried to make it clear to the medical team that he was still alive and that they needed to continue the CPR (Van Lommel et al. 2001: 2041). The truth is that the doctors discussed whether they would end the CPR or not but decided to give it one more chance. Also Sylvester's near-death experience resembles those described above. The elements of near-death experience and its frequency in Van Lommel's study were documented as follows.

	Elements of DNE	Frequency (n=62)
1	Awareness of being dead	31 (50 %)
2	Positive emotions	35 (56 %)
3	Out-of-body experience	15 (24 %)
4	Moving through a tunnel	19 (31 %)
5	Communication with light	14 (23 %)
6	Observation of colors	14 (23 %)
7	Observation of a celestial landscape	18 (29 %)
8	Meeting with the deceased	10 (32 %)
9	Life review	8 (13 %)
10	Presence of border	5 (8 %)

Source: Van Lommel et al. 2001: 2041

The study by Van Lommel and his colleagues and others include experiences of people belonging to other cultural settings. Furthermore, regardless of cultural setting, some were religious, others were not. Let us then take a look at near-death experiences of those belonging to non-Christian cultures.

A Jewish woman, named René Turner, was found unconscious after a car crash. At the hospital the doctor told her family that she had died. At that moment a frightened nurse ran toward them, telling the doctor and the family that René Turner was sitting upright in bed, asking not to be given any drugs, after which she sank into a deep coma. René recalled her near-death experience as follows: She said that she had no memory of dying or leaving her body but that she went through a dark maelstrom and that she was frightened. Then she saw a tiny dot of bright light that steadily grew brighter and came closer. She remembered feeling concerned for her mother, father, and sister. She arrived in an explosion of glorious light and found herself standing before a man in his thirties, about six feet tall, who had reddish brown shoulder-length hair and an incredibly neat, short beard and mustache. He was dressed in a simple white robe. Light seemed to emanate from him, and she said she felt he had great age and wisdom. He welcomed her with great love, tranquility, and peace but did not speak to her. She felt that she could sit at his feet forever and be content. He then showed her moments of her life. Not only did she relive these moments and realize what she had done and how she had hurt others; some of the things, she said, she would never have imagined could have caused any pain. At the moment she became burdened with guilt, she was directed to moments in her life that had given joy to others. She realized that there was balance between the two and that the balance was in her favor, and she received great love. She also reported having seen her grandfather, who looked younger than she remembered him. He told her that her grandmother would join him soon. This surprised her because her grandmother was in very good shape, travelling around the Continent. But he told René that her grandmother had cancer but did not yet know it. And indeed, her grandmother was diagnosed 3 months later and died of her illness. She then was told that she had to return, that she had a task to perform. She argued against it; she wanted to stay. While she glanced at her grandfather, all became black. The next thing she remembered was waking up in the hospital. René's recollection of her near-death experience also includes the most common features of NDEs, namely, the darkness (tunnel or space), moving toward the light, meeting deceased family and friends, and meeting a spiritual being. She did not have an out-of-body experience, but she had a review of her life.

In *Parting Visions*, Melvin Morse reports the following near-death experience of Mebruke, a 30-old Muslim woman. When she was about 20 years old, she visited Italy and swam in the Mediterranean Sea. When she became tired and headed for shore, she realized that she was too far away. She panicked, began to struggle, and swallowed water. Finally she slipped beneath the waves. She recalled that when she went under for the fourth time, her body went limp and that she was no longer aware of it. It was then that she saw a beautiful white light. It had such a calming effect that the more she looked at it, the more it calmed her. Until this day she is not sure what the light was, but she says that since Islam teaches about angels who are made out of pure white light, perhaps it was angels she had seen. When she was underwater, she heard a voice saying that she was not yet to die. Together with this voice came an energy shooting through her from her feet to her head, and, also simultaneously, she was propelled out of the water. It was, she recalled, as if someone was physically bringing her out of the water, but she can swear that no physical being was there. She was in a way *moved* through the water. Before long a boat reached her, and she was pulled out of the water.

It seems that near-death experiences do have cross-religious-relating features. Furthermore, some of the features also occur in near-death experiences of non-religious persons. At the same time, we recognize cultural aspects.

In conclusion, there are plenty of experiences of divine or ultimate reality. Regardless of the way in which such experiences are obtained (spontaneously, by way of meditation or prayer or at the threshold of death), regardless of the cultural-religious and personal settings of the experiencer, these experiences seem to have lasting effect on the lives of those undergoing them. The conversion can be phrased as – once I was blind but now I see; I was unconscious but now I am aware; I was asleep but now I am awake; I was lost but now I am found. The past is forgotten or changed, all things are made new. Suddenly those having undergone such experiences seem to know what to do; they have a purpose with and in their life.

The question is now, while experiences, regardless of them being religious ones or everyday experiences, are real for the experiencer, can they be justified?

Justifying Experiences

Thus far, I have argued that the sine qua non of (human) being (esse) is to experience. However, experiences are subjective or intersubjective at best. Hence, how can they be justified?

Since human beings are embedded in a natural-social-cultural-personal reality, they experience different levels of reality that constitute human reality. For many, this reality also includes divine or ultimate reality. Because of the embedment of reality in a natural-social-cultural-personal environment (through conceptualization of the experienced realities), the reality experienced is shared with others (including animals and perhaps vegetation). This means that even though there are discrepancies between experiences, human experiences are not, at least not always merely subjective.

However, some may ask whether it is, after all, not necessary in order for an experience to be justified that the experience is based on something independent of the experience? My answer to this question is, not necessarily. What is needed though is that there is something that *challenges* us, something that offers resistance (Herrmann 2004). The challenging something may be a thing, an event or a mental or physical process. To return to the experience of colors, it is the preconditions as explained by physics together with the human neurological conditions of perception that makes us experience colors. "Any experience is an expression of a relation between the experiencer and the experienced" (Runehov 2007: 43). Still, the question of justification is relevant, not least when the topic of the experience is not part of the material-physical world. Even though colors do not exist as we perceive them, the experience is justified by way of the physical pre-conditions related to neurological conditions. In other words, there is something that can be physically-neurologically grasped.

If we take the example of Shirley who experiences her arm as her pet rock detached from her body, her experience is justified due to a specific brain dysfunction caused by a brain-stroke. Knowing that such and such brain damage may cause such and such experiences justifies the experience. Both examples are justified because of reliable processes or faculties that underpin them. Nevertheless there is an important difference between the two examples.

The example of the experience of colors is subjectively justified but also intersubjectively justified. This is true even though there are people who have the condition of being colorblind. This cannot be said about Shirley's experience, which is subjectively justified (she actually experiences her arm detached from her body – she has a brain diagnosed as having a specific and correlating dysfunction). However, her experience is not intersubjectively justified. Even if her neurologist knows that her experience makes sense and is justified due to her condition, obviously, she will not agree with Shirley's maintaining that her arm is detached, because it is not and moreover, her arm is not a pet. What is commonly conceptualized as a pet is a cherished person, mostly a young person or child, and domestic animals.

Let us look somewhat closer at the justification of experiences. Keith Yandell divides experiences into those having *de facto evidence* and those having *effective evidence* (1993: 44). The idea is that an experience *x* is de facto evidence for a claim *c* if the experience *x* meets the relevant conditions (those conditions making the experience legitimated) which is, in my opinion sufficient to subjectively justify an experience (in Shirley's case, her neurological condition). An experience *x* is effective evidence if it is both de facto evidence and accepted as such by the experiencer. In this case, the experience is, again in my opinion, subjectively justified. Shirley accepts the experiences of her arm as her pet rock *x*. She cherishes it, hugs it as a pet. Hence, her experience is both de facto (neurologically) and effective evidence for her experiencing her arm as her pet (nobody can talk her out of it), which makes her experience subjectively justified. However, her experience can only remain subjective. In order to obtain intersubjective evidence or justification, she would need others to justify her arm as her pet, which obviously will not be the case.

What about the experience of colors? There are several conditions of color blindness. Total color blindness or visual agnosia, which implies that the person can distinguish colors but cannot perceive them. A more common color blind condition is red-green color blindness. A person having this condition has difficulties discriminating between red and green. Lastly, some have blue-yellow color blindness which has the consequence that the person has difficulties discriminating between blueish and greenish hues as well as between yellowish and reddish hues. This means that for some people, the experience of, for example the colors red and green, will neither have de facto (their visual condition does not allow them to perceive these colors) nor effective evidence (they cannot accept that they 'see' red and green). However, their experience of these colors as, for example, brownish, will receive intersubjective justification in the sense that there are many people who experience red and green as brownish but know that this is due to the condition of color blindness. They know that what they experience as brownish is experienced as red and green by the majority of the people. This means that an experience of red and green may not be subjectively justified, but an experience of red and green as brownish is both subjectively and intersubjectively justified.

Relating this to experiences of ultimate reality (including near death experiences), it also counts that the experiences are subjectively justified if they have de facto evidence and effective evidence. There can be different reasons why experiences of ultimate reality can be intersubjectively justified. Firstly, other people have experienced ultimate reality in a similar way (e.g. loss of notion of time, space and ego). Secondly the experience might simultaneously be shared with other people. Thirdly, one might find the necessary evidence in scriptures and other historical books, e.g. similar experiences have been undergone through history. Fourthly, one might find evidence in (clinical) reports of people having had similar experiences, e.g. out of body experiences at the threshold of death; etc.

Nevertheless, it might be the case that the experiencers do not recognize their experience as an experience of ultimate reality. In this case, as long as the experiencers do not recognize their experience as an experience of ultimate reality, and regardless of the amount of de facto evidence available through history, their experience of ultimate *x* cannot be said to be justified as an experience of ultimate reality. Indeed, sometimes it takes years before people realize what they have experienced and sometimes the person simply refuses to recognize the experience as such. Rather the person sees it as an entirely natural phenomenon. Hence, when it comes to experiences of divine or ultimate reality, belief seems to be essential. However, that belief is involved does not need to imply that the experiences (the item of belief) cannot be justified.

Justified belief is necessary for and closely linked to knowledge (Goldman 2000). Goldman's idea of justification is compared to act-utilitarianism in ethics, meaning that "an action is right if and only if it produces, or would produce, at least as much net happiness as any alternative open to the agent" (2000: 340; Runehov and Angel 2014: 216).

Hence, similarly to act-utilitarianism being a non-ethical theory, he wants a theory of justification that is non-epistemic. Goldman thus departs from traditional epistemic theories of justified beliefs. In other words, he is not interested in whether the believer knows that his belief is justified, or whether he can provide a range with evidence or arguments for his belief. Tout court, he is not interested in the justification by the person. Goldman argues that a justified belief gets its status of being justified from "*some processes* or *properties*", which according to him does not need to have an argument or reason (2000: 340; Runehov and Angel 2014: 216). If I have understood Goldman correctly, he would argue that de facto evidence, of whatever kind, is enough for a belief to be justified.

What is the difference between an epistemic theory of justified belief and a non-epistemic one? The difference lies in the modus operandi.

Modus Operandi for an Epistemic Theory of Justified Belief

If S feels *p* at *t* and S believes feeling *p* at *t*, than S belief of *p* is justified. In plain words, if you feel hungry at, say seven o'clock and you believe you feel hungry at seven o'clock then your belief that you are hungry is justified: i.e. you are hungry! Unfortunately, this way of justifying beliefs is not adaptable to the notion of credition, because there is no process, no dynamics, merely an observation, and a state of affairs. Now let us take a closer look at the modus operandi for a non-epistemic theory of justified belief.

Modus Operandi for a Non-epistemic Theory of Justified Belief

S is *ex-ante* justified in believing *p* at *t* if and only if there is a reliable belief-forming operation (process) available to S, which is such that if S applied that operation to his total cognitive state at *t*, S would believe *p* at *t*-plus Δ (for a suitably small Δ) and that belief would be *ex-past* justified. (Goldman 2000: 342, Runehov and Angel 2014: 217). The interesting point here is that Goldman's non-epistemic theory of justified belief is firstly, adaptable to both scientific and non-scientific beliefs and secondly, the distinction between ex-ante and past-ante and because of the addition of Δ data it is adaptable to the process of believing, i.e. credition.

To return to experiences of divine or ultimate reality, suppose a person has an experience of ultimate reality at 9 a.m. Then, within the person's mental state U, let S be the person, *p* be ultimate reality and *t* be 9 a.m. This means that the person is ex-ante justified in believing in ultimate reality at 9 a.m. if and only if the person has a reliable belief-forming process available to him/her (e.g. a revelation during a meditation session), which is such that if the person applies that process to his/her total cognitive state at 9 a.m., the person would believe in ultimate reality at 9 a.m. plus Δ (whatever happens during the day) and that belief would then be ex-past justified. This means that even if the person does not recognize his/her experience as an experience of ultimate reality, it still can by ex-past justified as such. However the person's belief in ultimate reality would not be justified in the classical epistemic way.

In conclusion, all experiences are subjectively ontologically real. An experience can only be epistemologically justified if the experience where S feels p at t, S believes feeling p at t. In other words the experience can only be epistemologically justified if there is both de facto and effective evidence. An experience can be non-epistemically justified if the experiences are ex-passed justified, or in other words, there is de facto evidence for the experience.

Conclusion

Human beings are experiencers and as such transcend different levels of reality, including ultimate reality. They conceive different types of reality during the course of their lives. They also share their experiences and create concepts and conceptions, which makes it possible to determine what the experiences are about. The advanced human brain sustains the richness and variety of human experiencing but also, the increasing richness and variety of human experiencing has causal effects on the brain. These facts have consequences for the cultural-religious-personal environment human beings are embedded in, for better and for worse. Experiences of ultimate reality are real in the same sense as are all human experiences. In other words, these experiences are subjectively ontologically real; they can be justified in an epistemic or non-epistemic way. However, that does still not imply that the object of experience necessarily *exists* independently of the experience. On the other hand, we can establish that when it comes to experiences of ultimate reality, the answer to whether such reality exists or not cannot be given.

Chapter 3
Human Uniqueness

Introduction

I have already mentioned that the human species differs from other animal species because of its advanced neocortex which enable humans to create language, art, myth and culture, and to make plans and create a self-image as well as a worldview. I also mentioned differences in DNA between the species. Humans and chimpanzees share 98.8 % of their DNA. Nevertheless, the species are very different in several ways. This is because two identical stretches of DNA can work very differently. Humans and chimpanzees use their genes in different ways, and even though the same genes are expressed in the same brain area in both species, they are expressed in different amounts. Hence already on a biological, neurological level, humans can be said to be unique, if we mean by unique that they have features no other animal species have.

The uniqueness of the human race has been discussed since the establishment of Darwin's theory of evolution announcing that the human race evolved from the same original cell as all other animals. Humans are not the 'crown' of creation, there is no crown and no creation. Rather, biological principles such as randomness, adaption and natural selection led to the evolution of different species including the human species. This line of thinking has not only put the human species on the same biological scale as other animals; it also tries to put the human race on the same social scale. For instance, honoring zoologist Donald R. Griffin (1915–2003), Carolyn A. Ristau compiled a volume on non-human animal cognition, because Griffin had challenged the behaviorist's claims that animal communication is characterized as merely *groans of pain.* (Ristau 1991: 14). Paradoxically, Ristau's idea that non-human animals might be aware of their own thoughts triggered a new way of thinking about the uniqueness of the human animals. Questions are raised such as, are non-human animals also social animals or does the human social ontology remain unique? What external and/or internal information processes underpin such

A.L.C. Runehov, *The Human Being, the World and God*,
DOI 10.1007/978-3-319-44392-8_3

social reality? Different research results suggest that social facts are common, constant and universal, that the same biological laws are integrated in all parts of nature and that, because social facts are subject to these laws, social facts are integrated in all parts of nature as well. In addition, these social facts improve from generation to generation. The question remains though, to what degree are the social facts integrated in different parts of nature, or more precisely, in the lives of different animal species? I will argue that

1. All animals are embedded in a social system.
2. However, that does not mean that all animals are social animals in a true sense.
3. That the human social ontology remains unique.

Social Animals

Criteria for Being a Social Animal

What are the criteria for being a social animal in the true sense? Inspired by Raimo Tuomela's *Social Ontology* (2013), I suggest the following criteria to be necessary. *Firstly*, one needs to possess a self-identity as well as a group-identity and a self-identity within a group. Possessing self- and group-identity requires self-recognition and self-recognition within a group. Only then is it possible to have a cognitive sense of I and me as well as I and we and also of I and me within a 'we'. *Secondly*, one needs to understand intentionality as well as understanding collective intentionality something which requires possessing the Theory of Mind. To put it differently, one has to *know* one's own state of mind in order that one can infer the state of mind of others from one's own. Collective intentionality also requires mutual responsiveness and a form of coordination between the members involved. Furthermore, the members need to be sensitive to each other's behavior and they need to understand each other's action as a specific intentional action which they are able to respond to. *Thirdly*, the members of the group have to be committed to the joint action.

From this I argue that the criteria necessary for being a social animal is represented in different degrees within the animal kingdom and human beings. For example, while all animals possess some social skills, only few possess a degree of self and group identity and understand intentionality and group intentionality, e.g. the great apes and (bottlenose) dolphins.

However, I also argue that even if some non-human animals may fulfill the criteria for being a social animal it still is the case that the human social ontology is unique. Perhaps the human species is not the crown of creation but the human race has the genetic and social powers to rule creation, for better or for worse.

Are All Animals Social Animals?

The Greek philosopher Empedocles (495–435 BCE)[1] argued that animals co-operate with each other (in Allee 1931: 386). More than 2000 years later, the French philosopher Alfred Victor Espinas (1844–1922), was convinced that *all* animals were indeed embedded in some social environment at some point in their lives. According to him, "[n]o living being is solitary. [Not even those who are] devoid of distinct and separate sexes [...]" (Espinas 1878 in Allee 1931: 386). If it would have been otherwise, conservation and renewal of life would not have been possible. Allee writes:

> Communal life, therefore, is not an accidental fact in the animal kingdom; it does not arise here and there fortuitously and, as it were, capriciously; it is not, as is so often supposed, the privilege of certain isolated species in the zoological scale, [...] but, on the contrary [...] a normal, constant, universal fact (1931: 387).

Hence, a social life not only seems to be a common feature of life, it is also a necessary one. Social facts are subject to biological laws that are the same everywhere and constitute a "homogeneous whole thoroughly integrated in all [nature's] parts" (Allee 1931: 387).

We can reconstruct this as

1. Social facts are common, constant and universal
2. The same biological laws are integrated in all parts of nature
3. Social facts are subject to these laws
4. Hence, social facts are integrated in all parts of nature.

Espinas (1878) also observed a development of the social habits. Indeed, not only did each generation pass on their social skills to the next generation, they also added at least some improvements to it. We can also add a fifth item:

1. Social facts improve from generation to generation

Hence, the social facts together with the biological laws combine to establish an increasingly complex intertwined system allowing novel social features to emerge.

Entomologist William Morton Wheeler (1865–1935) followed a similar line of thought when he maintained that not only do living beings struggle and compete for food, mating and shelter, which are all issues of *primary and individual* survival, they also co-operate in order to further develop their social structures, which are issues of *secondary and group survival* (Wheeler 1928).

Furthermore, according to Warden Clyde Allee (1885–1955),[2] the first steps towards development of societies were already taken when life began on our

[1] Empedocles, see Audi 1999: 261–262.

[2] The Allee effect is a biological theory stating that there is a positive correlation between population density and the per capita population growth rate in very small populations. Simply put, for very large populations, the reproduction and survival rates of individuals decrease with population density. This contrasts with smaller populations, where less population density slows the growth rate of the population, due to intra-specific competition.

planet. His argument is based on his observations that even loosely organized as well as unorganized groups of animals living temporarily in the same region will co-operate for the sake of survival (1931). In other words, animals may congregate in some specific region at some specific time because of the survival value of that particular place at that particular time. Once the animals become adjusted to their new environment and become stronger, the survival value of the group may no longer be obvious, until existence gets less favorable again. For example, when the place becomes too crowded causing a higher rate of death amongst the members of the group, individuals and smaller groups will leave the place, in search for a new and more promising territory to rebuild a social system or die out.

Thus far, all this is not very different from modern human communication and societies. Just think about the phenomenon of urbanization. Because of poor life circumstances in the rural areas, people moved to cities. However, since the 1950's, at least in the richer parts of the world, we have experienced a counter-urbanization. Indeed, people are moving away from cities. There are several reasons for this behavior but one reason is that cities have become more and more unhealthy places to live due to pollution, crime and traffic congestion. Another important reason is the development of information technology (passed on to and further developed from generation to generation). Not only have new business parks emerged outside cities, today's communication facilities also allow us to connect with the rest of the world from wherever we are situated.

I believe it to be correct to argue that all animals are indeed at some point in their lives embedded in a social system, but, can it be argued that all animals are social animals? What does being a social animal require? At least two mental traits are needed, namely identity and intentionality. Let's start with identity.

Self-Identity and Group-Identity

To be a social animal in the real sense, the animal needs to be able to form a self-identity as well as a group-identity. In other words, the animal needs to be able to have a *sense* of *I and me* and a sense of *we and I within us*. Three distinctions can be made.

1. Animals that are part of a social group but are part of it as individuals. They may have their specific places within the group. For example, they may be the leader of that group. Perhaps one could argue that these animals have a sense of 'I' that has a specific task within a 'we'. However, when threatened, their individual survival instinct will prevail. Another way to put it is that they are *more self-centered than group-centered.* Most mammals belong to this group.
2. Animals having a group-identity but no or little self-identity. These animals are *group-centered rather than self-centered.* Ants belong to this group of animals.

Due to this, the behavior of ants is often used as an example of biological altruism.[3] Indeed, ants act as if they were one biological system that consists of different 'modules' each performing their duty to fulfill the collective task. They act as neurons act in the brain.

3. Animals that have *both self- and group identity*. They are able to think in an *I-we-mode* as well as in a *we-mode that includes the I-mode*. Some mammals (the great apes and dolphins) and humans belong to this third group.

The animals of the first and second group have their specific social systems but, could it be argued that they are social animals in the true sense? Even though the animals of the first group may have a sense of identity, do they know that they are an 'I' and a 'me' and recognize this 'I' and 'me' within the group? In other words, do these animals possess the ability of self-recognition?

Humans recognize themselves in a mirror from an early stage of infancy. However, there is evidence that the great apes (common chimpanzees) as well as bottlenose dolphins also possess this ability (these belong to the non-human animals of the third group). Experiments suggest that both species behave in a way that it seems like they investigate the marks put on their body in mirrors or other reflective surfaces put around them (Reiss and Marino 2001: 5942; Reiss 2011: 155–157). The Dolphins that were marked spend more time in front of the mirror compared to those who were not marked. They even selected the best reflective surface available to view their markings. Like the great apes, dolphins seem to think, *I have never seen these marks on my body before – let's investigate them.* After a while, *hmm! These marks do not look that bad and they do not hurt me – OK – I'm fine*. Frans de Waal and Joshua Plotnik performed a similar study with elephants. One of the elephants passed to *mark test* by using its trunk to pat the mark placed on its head (in Reiss 2011: 186). But Diane Reiss was not yet entirely satisfied until she, in another study observed that the dolphins

> would look at the inside of their mouths, often holding their mouths open wide in prolonged gapes, they gazed at their eyes, holding their eyes right near the mirror surface; they blew bubbles at the mirror; they twisted and turned their bodies in bizarre postures [...] and they brought their toys to the mirror and engaged in toy play there (Reiss 2011: 158).

Hence the dolphins engaged in self-directed behavior, similar to children who, for example pick their noses, make funny faces and play in front of a mirror. Now she was convinced that the dolphins understood that the external image in front of them, the dolphins in the mirror, were themselves. They possess self-recognition.

The important issue is that, because animals that are able to recognize themselves in a mirror also recognize the other members of a group as similar to but separated from themselves; it can be argued that these animals possess self- and

[3] Biological altruism if found in the sterile workers in the social insect colonies (ants, wasps, bees and termites). These workers devote their whole lives to the queen, constructing and protecting the nest, gather in the food, etc. According to Samir Okasha, such behavior is the most altruistic, (though biological,) because the workers, since they do not leave any offspring of their own, have zero personal fitness. They devote their whole lives to increase the reproductive possibilities of the queen (Okasha 2008).

group-identity, which are two criteria of being a social animal in a true sense. Let us now look at the other mental traits that the animals need to understand, namely intentionality and collective intentionality.

Intentionality and Collective Intentionality

To be a social animal in a true sense, the animal needs to understand, to be able to explain (not necessarily by way of language) as well as to predict its own emotional and cognitive behavior and that of others not only from its own point of view but also from the point of view of the social group.

As John Searle argues, intentionality not only includes intending, as in "'I intend to go to the movies', but also includes beliefs, hopes, desires, emotions, perceptions, and lots of others" (2008: 31). Similarly, group or collective intentionality can be described as we intend, we believe, we desire or, it is the group's (our) intention, belief, desire ... This means that the animal needs to possess higher-order thought. The animal's belief that *p* needs to include a representation not only of *p* but of the animal's belief that *p*.

Daniel Dennett refers to this ability as possessing the concept of intentional ascent (1987). In other words, the animal is not only aware of its self-directed intentional states but also of the other's directed intentional states.

Andrew Whiten (2000) added an interesting distinction, namely between a mental representation of a mental representation on the one hand and a mental representation of a mental representation as a mental representation, on the other hand (in Browne 2004: 649). The animals have to be aware of their own states of mind so that they can infer the states of mind of others from their own. As such they put themselves mentally in the other animal's mental state. Once again, there are experimental studies suggesting that the great apes and the bottlenose dolphins are at least aware of their own state of mind.

Derek Browne performed such a study. He showed that bottlenose dolphins are actually able to not only discriminate between higher and lower tones, but they are also able to choose a middle path, i.e. the path 'I do not know whether I hear a high or a low tone' (2004: 641). At least it seems that the dolphins 'know' that they 'do not know'.[4]

The question is however, whether the dolphins possess meta-cognition or are we rather talking about first-order cognitive responses? In other words, is the dolphin really aware that it is in this particular mental state (in this case the state of uncertainty) or does it feel uncertain or conflicted and somehow this feeling directs it to the escape paddle? (Browne, 2004: 650).

To settle the matter, another research group, David Smith and his team, performed experiments with humans, dolphins and great apes. Surprisingly or not, the

[4] It has to be mentioned that, while they were trained to hear high versus low sounds which means that there were stimuli involved in training them, the escape paddle was not under any stimulus!

results showed that the response patterns were similar, except for the fact that humans tended to use the escape paddle less often (20 % against 45 %). According to Smith et al., because the dolphins, the chimps and the humans responded similarly to a complex pattern and because it is known that humans use their meta-cognitive abilities, they concluded that the great apes and the dolphins also need to use meta-cognitive abilities (1995).

We can reconstruct this argument:

1. Human beings possess meta-cognition.
2. Meta-cognition implies responding to a complex pattern in a particular way.
3. Human beings, great apes and dolphins respond in a similar way to complex patterns.
4. Therefore, great apes and dolphins also possess meta-cognition.

Of course this line of thinking is valid but is it also sound?[5] In my view, there are some problems with premises 2 and 3. Firstly, meta-cognition not only implies that one responds to complex patterns in a particular way, the main definition of meta-cognition is knowing about knowing, knowing one's own cognitive processes. As mentioned earlier, even though there were no stimuli involved in the experiment with the third paddle, it does not provide clear evidence that the dolphins really were aware of their cognitive (mental) state of being uncertain. Secondly, from the fact that humans use their meta-cognition (theory of mind) to perform a task it cannot be deduced that dolphins and chimps also possess this capacity on the same level as humans. Nevertheless, there is at least some evidence that some non-human animals possess self- and group-identity and that they may understand intentionality. Hence, let us continue with investigating collective intentionality.

Collective Intentionality

According to Searle, collective intentionality is a genuine biological phenomenon (2008: 31). He means that, a social reality (or fact) is established as soon as two or more humans or animals are involved (2008: 32). If we accept his view, all animals should be capable of establishing social realities. However, in my opinion a distinction should be made between biological intentional behavior and agent-controlled intentional behavior.

As Fred Dretske also argues "to be an agent it is not enough to be a thinker and a doer. The thinking must explain the doing" (Dretske 1999: 19). Only then, he argues, can we talk about expressions of agency (Dretske 1999: 20). Only then we

[5] For those who are not familiar with analytic philosophy: An argument is valid if it is deductively valid. This is, if the premises necessarily imply the conclusion. However, a valid argument does not need to have true premises. An argument is sound if it is valid, noncircular and contains only true premises. If an argument is sound, it means that the conclusion is not only necessarily implied by the premises but also necessary true.

can talk about agent-controlled intentional behavior that besides the I- and we-mode is a necessary condition for understanding collective intentionality and for being a social animal in the real sense. For instance, all hungry animals intend to find food, for themselves and during a certain period also for their offspring. This type of intentionality is what is meant by biological intentionality or, in Dretske's words, animal behavior. However, only some animals and humans may have the intention of finding food for the social group they are part of and only some animals and humans may have the intention to find food for members of another social group. This type of intentionality fulfills the criteria of agent-controlled intentionality or in Dretske's words, agent action (Dretske 1999: 19–31).

Finding food for yourself and your offspring belongs to natural behavior; it is an instinctive and evolutionary basic. Finding food for other members of your society and members of foreign societies may enhance your own possibilities of survival but your actions need careful planning and strategies. Take for example, carefully planning investments in undeveloped societies with the aim to increase local employment and income thereby improving the environmental quality, which in turn improves the environmental quality of the whole world to which one belongs. Humans know what helping others means for the others and for themselves; their intentions are agent-controlled.

However, once again there is evidence of non-human animals helping members of other social groups. Indeed, bottlenose dolphins seem to behave as intentional agents (Pryor and Shallenberger 1998). These dolphins have been fishing alongside humans at Laguna (Brazil) since 1847. They actually herd fish towards the fishermen who wait in the shallows. Once the fish are close to the fishermen, they signal the fishermen to cast their nets. Furthermore, they go after the fish that escape the nets. According to Pryor and Shallenberger, this particular behavior has not been found in other dolphin populations (1998). Apparently, this particular intentional act has some specific meaning for the Laguna bottlenose dolphins. However, on the basis of this, can we conclude that these dolphins that seem to act as agents also are agents? In other words, do they understand that they are giving a signal to the fishermen to cast their nets or have they acquired this unique signal through blind, mindless, year-long trial and error learning? (Pryor and Shallenberger 1998: 161–199).

Yet other questions are, firstly, can we conclude that the dolphins help the fishermen, because they know what helping them means? Secondly, is their intentional behavior biological intentional or agent controlled? For us to be able to answer these questions we need to make yet another distinction. In Dretske's words, "there is a difference between being caused by an event that means M and being explained by the fact that it means M. (Dretske 1999: 20). In the case of the dolphins, their helping the fishermen (the event) caused the fishermen to sell fish on the market and buy other necessities (M). However, this does not mean that their selling fish and buying other necessities is explained by the behavior of the dolphins. To put it differently, it is doubtful that the dolphins, in contrast to a human animal could form a belief or predict the intention of the fishermen. The reason is that if they can, they need to be able to infer the fishermen's intended actions from their own repertoire. In my opinion, it is hard to see that they would be able to do so; dolphins do not sell or buy things, so how could they predict that driving in the fish to the fishermen would help the fishermen to sell the fish?

But there is more to the story of collective intentionality. As mentioned earlier, the following criteria need also to be fulfilled:

1. Mutual responsiveness.
2. A form of coordination between the members.
3. Sensitivity to each other's behavior.
4. Understanding of each other's action as a specific intentional action which one is able to respond to.
5. Commitment to the joint action.

The question is now; to what extent are these criteria fulfilled by non-human compared to human animals (Tummolini et al. 2006: 117)?

Non-human animals do instinctively capture the intention of the group to collectively, say, escape from some danger. Practically, when hearing a certain sound of one of the animals in the group, animals A, B and C will start to run. In other words, they will collectively behave in a certain way. The act is intentional in the sense that A, B and C intend to escape danger. There seems to be a mutual responsiveness (to the specific sound) and some form of coordination between A, B and C's behavior (running together in the same direction). Hence criteria 1 and 2 may be met. However, could it be argued that A, B and C are sensitive to each other's behavior, that they understand each other's action as a specific intentional action to which they are able to respond? Furthermore, could it be argued that animals A, B and C are committed to their collective act?

In my opinion, that the latter criteria are fulfilled is not clear. As Searle puts it: "[c]ollective intentional behavior […] is not the same as the summation of individual intentional behavior" (2010: 42–61). In our example, the intentional behavior of animals A, B and C running for their lives does not necessarily represent collective intentional action but does represent a summation of animals A, B and C's individual intentional behavior. What about the great apes and dolphins?

If we return to the Laguna dolphins, one would like to say that, indeed all criteria are fulfilled. However, we need to remember that this complex strategic behavior of the dolphins has been going on for more than 100 years – we cannot exclude the importance of training, learning, trial and error.

Another study performed by Michael Tomassello and his team suggests that dolphins do share intention and that this "might be one end of a continuum that also contains 'awareness of others' and 'awareness of other's intentions', as well as the more basic ability to form an intentional plan of action" (2005: 706). While he was snorkeling around three wild rough-toothed dolphins (two adults and one youngster) he saw how one of the adult dolphins trailed a piece of plastic from one of its pectoral fins. Then the three dolphins passed this plastic piece back and forth to one another. They did not make any attempt to steal the plastic but waited for the plastic to be released. They also passed the plastic on to the youngster, seemingly to ensure the participation of the young dolphin. This observation of wild dolphins, in their own free environment, suggests that dolphins do have the capacity of collective intentionality. Their play is not caused by training or learning, trial and error. It comes spontaneously just as play comes naturally to children. Tomassello et al. concluded that such behavior might provide the basis of simple forms of culture (2005: 706).

Perhaps we may suggest that some animals are or at least may be social animals in a similar sense as human animals are. The experiments mentioned clearly suggest the probability that some non-human animals fulfill the criteria for possessing self- and group identity, intentionality and collective intentionality. However, there still is something unique about human social ontology. While some non-human animals may be cultural in a basic way, human culture is far more complex. Humans are also institutional animals.

Institutional Animals

Humans are not only social animals; they are also cultural or institutional animals (Runehov 2012a). Humans not only live in societies, they create their societies; they create culture and institutions. Take the example of trade.

In the early times of human social evolution, the form of trade that was used was barter, i.e. people simply exchanged goods and services. Later, barter was replaced by commodity, money, i.e. any commonly available commodity that had intrinsic value (seashells, cacao beans and even bread). From about 2000 BC currency was introduced as a medium for trade. In the beginning, the currency was represented by standardized coins, later paper-money was introduced and today trade is mostly done by electronic transactions, i.e. by cyber-money. The point is that money, being a piece of paper or an amount of numbers on a computer, is collectively accepted to have what Searle calls *deontic power.* Money, unlike a candle, performs a certain function not in virtue of its physical structure but in virtue of its collective attitudes (Searle 2008: 33). In the language of logics: X counts as Y in context C. Other examples are wedding rings, passports, certificates, stock-markets, contracts, corporations, etc. (Searle 2008: 39). Only human social ontology includes deontic powers. Non-human animals do not need contracts stating that they have the right to live in their 'house'. The question is why did human social ontology become so advanced? I suggest there are three main basic reasons.

The Underpinning Evolutionary Mechanisms

The first reason has already been emphasized by Charles Darwin in his theory of evolution. He asserts that species are *populations of individuals* that carry a pool of genetically acquired information through time. Furthermore, political and social complexity is driven by population growth. Surely, this is something which is mirrored in the human societies. (Richerson and Boyd 2006: 59). However, do not all species carry such a pool of information through time? Still, we only find political and social complexity amongst the human species. Why? I think it is so because of the second reason.

The second reason why human social ontology became that complex may be the idea that a *co-evolution of genes and culture* took place, which is more than a passing-on of social skills from generation to generation. Co-evolution can be explained as evolutionary systems "in which two species are important parts of each other's environments so that evolutionary changes in one species induce evolutionary modifications in the other" (Richerson and Boyd 2006: 192).[6]

The gene-culture co-evolution played a crucial role in genetic evolution of human psychology because genetically evolved psychological biases steer cultural evolution in genetic fitness enhancing directions. On the other hand, culturally evolved traits affect the relative fitness of different genotypes in many ways. Take for example a culture in which punishment is adapted. Indeed, culturally evolved moral norms can affect fitness if the violators of these norms are punished. The violator gets the chance to reevaluate the actions (enhancing fitness) or experiences a decrease of social fitness (imprisoned, not respected, becoming an outsider). (Richerson and Boyd 2006: 193). Richerson and Boyd argue that genes and culture are obligate mutualists.

Genes alone cannot promptly adapt to the rapid environmental changes and culture would not exist without brains and bodies. Hence, genes and culture are tied together but culture remains nevertheless subject to evolutionary forces that pull behavior in different directions (Richerson and Boyd 2006: 194). This line of thought seems to go hand and hand with Allee's, namely that social facts are subject to biological laws. The difference is that in Richerson and Boyd's view, there is strict mutual causation between the two.

A third reason for the uniqueness of the human social reality is explained by way of the *social complexity hypothesis*, a key driver of human intelligence. This hypothesis means that animals living in large social groups (as apparently humans do) should display enhanced cognitive abilities. The larger a society becomes, the more social cognition is demanded.

This makes sense, because within each society, the individuals have to recognize the other individuals; they continuously need to keep track of the social position, social behavior and social success of other individuals; they have to classify the others by age, gender, genetic relationship, family status and social rank, physical capacities, reliability, preferences, expertise, etc., and update all this information as circumstances change (Byrne and Whiten 1988). Moreover, because humans typically align themselves with different sub- and side-groups, (e.g. religious and

[6]The term co-evolution was coined by Paul Ehrlich and Peter Raven in 1964 in order to describe the evolutionary relationship between butterflies and plants. Caterpillars prey on plants, and the plants in turn evolve chemical defenses to mitigate the damage of insect attack, which leads to the evolution of caterpillar detoxification. Since then its meaning has been extended to any case in which two distinct evolutionary systems interact in interesting ways (Richerson and Boyd 2006: 276).

political groups) they must understand and remember how each member relates to these sub- and side-groups and how the different groups relate to each other. The more socially differentiated and complex groups become, the more cognitive power is needed. Hence, at some point in the history of evolution, the neocortex simply had to evolve and further develop.

To conclude this section, all species are populations of individuals that carry a pool of genetically acquired information through time. Social facts, since they are subject to biological laws that are integrated in all parts of nature, are also integrated in all parts of nature and furthermore the social facts improve from generation to generation. There is evidence that some non-human animals are or may be social animals in a true sense. However, because the gene-culture evolution only took place in humans, humans not only live in societies, they create societies and complex cultures and institutions. Humans acquired a gene pool (neural system) that enabled them to not only reproduce but also to enhance the fitness of themselves and their offspring. Therefore, they could establish large social groups including sub- and side-groups. Also, the cultural traits influenced the human gene-pool and vice versa, which in favorable circumstances, enhanced their cognitive abilities further and further, which in turn gave rise to more and more complex social systems. It is this that remains unique for the human-animal.

Artificial Intelligence

Compared to non-human animals, humans seem to have specific features unique to the species. Amongst other things, these specific features have inspired computer scientists to develop human-like devices. The human brain was studied in order to build 'thinking machines' by using artificial intelligence (AI). The term AI was coined by computer and cognitive scientist John McCarthy (1921–2011) in 1955. According to him, "[m]achines as simple as thermostats can be said to have beliefs, and having beliefs seems to be a characteristic of most machines capable of problem solving performance" (McCarthy 1979). The main idea of AI is to simulate features of human intelligence such as reasoning, knowledge and classification, which are the main features. Furthermore, the idea is to simulate planning, learning, communication, perception, and manipulation of objects. Obviously the research field is interdisciplinary; besides computer science, the field includes disciplines such as mathematics, psychology, linguistics, philosophy and neuroscience. That machines simulate human intelligence raises several philosophical problems, some of which will be discussed below. Furthermore, AI has been and still is a source of inspiration for science fiction novels and movies. This section discusses two philosophical problems.

The first problem concerns whether a computer can think, as compared to a human being's thinking. The second problem concerns whether a computer can feel, as compared to a human being's feeling. The final question is how human are androids?

Thinking Computers

Alan Matheson Turing was first to illustrate a challenge for humanity. He did this by way of an experiment, which is known as *Turing's test* or the *imitation game.*

Roughly, two people, Lisa and John, are placed before a computer. Both are sitting in their own cubicle and only have contact with the world outside by way of a terminal on which they write the answers to the questions of the experiment. The task of the experiment leader is to find out which answers are Lisa's and which are John's. Lisa has to answer the questions in a way that confuses the experiment leader while John is supposed to answer the experiment leader in a helpful manner. After a while, computer Z112 programmed to simulate human behavior replaces John. If the answers coming from Z112 are in such a way that the experiment leader takes them to be John's, the computer is considered to be able to think, i.e. to behave with human intelligence, according to Turing (Turing 1990: 40–66).[7]

In the 1950s when Artificial Intelligence (AI) was born, the main goal was to exhaustively copy human cognition and language on a digital computer. According to Terry Winograd, this goal has two distinct aims: firstly, to explain human mental processes as thoroughly and unambiguously as possible, and secondly to create intelligent tools, not necessarily duplicated from human intelligence, applying intelligence to serve specific purposes (Winograd 1991: 201). The problem-solving power of an AI program derives from the knowledge it possesses (Winograd 1991: 208). However, we need to observe that knowledge is seen as a kind of commodity to be produced, refined and packaged; there is no concern with the epistemological problems of what constitutes knowledge and understanding.

Nevertheless, defenders of AI argue that the methods and theoretical foundations that are thus far applied may be extended to cover the full range of cognitive phenomena, including human knowledge and understanding (Winograd 1991: 209). I will return to this later.

Philosophical Opposition

This was the time for philosophers to wake up. How could a machine be compared to a human being? Could a machine possibly think? There were many philosophical proponents but the one who is important for the present discussion and whom I hence want to mention is John Searle.

He argued that machines cannot think since they lack consciousness, they lack intentionality (1983). To illustrate his objection to the suggestion that a machine could possess cognitive capacity similar to a person he put forward the *Chinese room* thought experiment (Searle 1980). Roughly, the idea goes as follows. Assume Searle is in his office. The door is locked from the outside. Searle does not know any

[7] See also, Runehov 2008.

other language than English. Through ticket windows, he and Chinese people exchange notes consisting of Chinese characters. When he receives a note, Searle has to compare the written character with the signs in an instruction manual he has at his disposal. This manual is written in English and explains how he has to compose a Chinese answer to the signs on the note. However, the manual does not give any guidance concerning the meaning of the sign.

For example, if he receives a note with the Chinese character 扶, he has to answer it with 手. To Searle the characters 扶 and 手 are completely meaningless symbols, but not for the Chinese persons. Searle in the Chinese room is suggested to be analogous to a computer program, i.e. both perform a pure syntax manipulation of symbols. Searle does not have any clue what the Chinese signs mean, he simply does not understand Chinese. Nevertheless, the Chinese room experiment would pass the Turing test.

However, what Searle wanted to show is that to manipulate symbols is not sufficient to understand them in a cognitive sense. Searle's thought experiment wanted to show that computer programs are not able to think, i.e. to behave with human intelligence (Searle 1990).

The first problem with Searle's thinking is a conceptual one. As often, within the sciences, concepts are differently understood. In the language of computer science, the concept of intelligence is understood as follows:

> an entity is intelligent if it has an adequate model of the world (including the intellectual world of mathematics, understanding of its own goals and other mental processes), if it is clever enough to answer a wide variety of questions on the basis of this model, if it can get additional information from the external world when required, and can perform such tasks in the external world as its goals demand and its physical abilities permit (McCarthy and Hayes 1969: 4).

In other words, computer scientists define intelligence as a practical unit. This definition of intelligence includes an epistemological and a heuristic part. The former part is understood as a representation of the world in a way that problem solving follows from the information given in the representation. The latter part, on the basis of the information given, solves the problem and decides what to do. (McCarthy and Hayes 1969: 5). In his article on the concepts of information, knowledge and intelligence, Alistair MacFarlane, sees intelligence not as a practical unit but as a process or an innate capacity to use information in an adequate way. He writes,

> It is a capacity to acquire, adapt, modify, extend and use information in order to solve problems. Therefore, intelligence is the ability to cope with unpredictable circumstances. But intelligence is not merely analytical: to survive and flourish in society, we must also have social and emotional intelligence. Intelligence involves a capacity to adapt and learn that has been accumulated throughout our species' development, yet intelligence is something we have by virtue of our individual genetic inheritance. By assiduous practice we can become more skilled. By diligent study and careful observation we can become more knowledgeable. But there is no way we can personally become significantly more intelligent. You can no more double your natural intelligence than you can double your natural height (2013).

Hence, intelligence is connected to the person's biology.[8] Intelligence is not a skill but an innate capacity to acquire skills. With this definition of intelligence and in order to have artificial intelligence similar to human intelligence, one would need a system that possesses some kind of genes. Since there are no computers or robots with genes, they cannot be intelligent. These devices remain symbol manipulating machines. Concerning the subject matter of intelligence, computer scientists and philosophers do not agree.

One of the objections to Searle's Chinese room argument and which is of interest for our discussion is the *Robot objection.* In the robot argument, it was suggested to replace the computer by a robot with arms, legs and a camera. It was then argued that the robot as a whole might behave with human intelligence, because, to interact with the environment, to have impressions, to move and to handle things are essential parts of the process of thinking and understanding. Cognitive capacity, it was maintained, also includes perception and motoric abilities. Hence, cognition is not only a matter of symbol manipulation, because it also comprises of a range of causal interactions with the environment.

With respect to the capacities these robots have, that they are built with copies of human physiology like eyes and arms, does not make them biological beings. Searle is still correct when he counter argues that a robot merely manipulates patterns without knowing what these patterns actually mean.

Winograd compares Artificial Intelligence (AI) with bureaucracy and writes: "The techniques of AI are to the mind what bureaucracy is to human social interaction" (1991: 211). AI has similar advantages as bureaucracy, for example its rational character, a lot of rules, means-end calculations and elimination of personal commitment or responsibility (Winograd 1991: 214).

To conclude this far, the fact that computers use artificial intelligence does not imply that these computers are thinking machines, i.e. intelligent in the same sense as humans are. Because comparing human intelligence with artificial intelligence is not fruitful, I suggest we leave this discussion as it is. What remains an interesting question is to what extent artificial beings copy human beings.

Neural Networks

Trying to avoid the practical problems with artificial intelligence, computer scientists began to consider the possibilities of neural networks. Computational neural networks are said to simulate the brain's network structure.

Briefly, the data is distributed and stored as a pattern in the connections (weights) between the different nodes (neurons) of a neural network. Each node operates on simple general principles. The basic idea guiding this work is that cognitive

---

[8] It is not my intention to dig deeper into the philosophical debate concerning intelligence. The point I want to make is that whether AI can be regarded to have intelligence depends on how intelligence is understood.

structures in organisms emerge through learning and experience, not through explicit representations and programming. There are no rules for how problems are to be solved. Instead, the rules describe how the nodes are connected to each other, how the weights are to be modified and how the information between the nodes is transferred. A learning process is activated by presenting different patterns as well as correct answers to the program. Neural networks are now successfully applied in computer programs for picture analysis, recognition of fingerprints, speech and face recognition, interpretation and translation of written text, weather forecasting, stock-marked prognoses, social media, and so forth. In fact, neural networks are part of our lives to such an extent that we can no longer do without them. However, that it is possible to simulate the human neural system does not mean that a computer neural network has become equal to the human neural system.

Humanoids, Androids and Humans

Robotics is an interdisciplinary endeavor including engineering, cognitive science and neuroscience, linguistics, psychology and other for the task needed disciplines. A humanoid robot is a robot of which its overall appearance is based on that of the human body. The humanoid is able to interact with the human environment and is able to use human tools. In general humanoid robots have a torso with a head, two arms and two legs. However, some humanoid robots only model a part of the body, for example, from the waist up. Some humanoid robots also have a face. They have eyes, ears, nose and mouth and can be quite beautiful.

Androids are humanoid robots built to aesthetically resemble a human.[9] While humanoids do not need to be robots, it suffices that they resemble human beings, androids are robots. The main difference between a humanoid robot or an android and other types of robots (e.g. industrial robots) is that it can adapt to changes in its environment and it can change itself in order to reach its goal. Generally humanoid robots and androids are more complex than other kinds of robots. Humanoid robots and androids possess human characteristics such as motivation, emotions and cognition. In order to construct these humanoid robots scientists study, amongst other things, the relationship between parents and children, not least attachment theory.

Psychiatrist John Bowlby, perhaps the pioneer of attachment theory, did not approve too much of Freud's idea of a superego that has its origin in the resolution of the Oedipus complex. Instead Bowlby argues that during the early years, while the child acquires the capacity for self-regulation, the mother is her child's ego and superego, he writes,

> It is not surprising that during infancy and early childhood these functions are either not operating at all or are doing so most imperfectly. During this phase of life, the child is therefore dependent on his mother performing them for him. She orients him in space and

[9] Androids see, http://roboticsandtechnology.blogspot.se/p/humanoid-robot.html, accessed Wednesday, September 30, 2015.

> time, provides his environment, permits the satisfaction of some impulses, restricts others. She is his ego and his super-ego. Gradually he learns these arts himself, and as he does, the skilled parent transfers the roles to him. This is a slow, subtle and continuous process, beginning when he first learns to walk and feed himself, and not ending completely until maturity is reached....Ego and super-ego development are thus inextricably bound up with the child's primary human relationships (1951: 53).

The attachment theory concerns the relation between a mother and her child. It especially concerns an inter-reaction where the child prefers to be close to the mother in unfamiliar situations, especially when young. It concerns what happens with a child who experiences a lack of attachment with its mother. It is by studying such unfortunate children that psychologists and psychiatrists know how important a mother-child attachment really is.

Robotic science, in order to create human like robots realized the importance of this theory. After all, the aim of constructing humanoid robots is that they will easily mingle with humans and function as humans in a human environment. In the language of robotic science the attachment theory is called *distributed cognition* and is similar to the behavior-based system in the sense that both methods are used to understand intelligence by way of human-human and human-robot interactions (Ishiguro 2005: 1).

Distributed cognition is vital for, e.g. companion and entertainment robots. These humanoids and especially androids since these robots are copies of existing human beings, give humans the illusion of them being alive and furthermore that an emotional bonding can be established between the human and the robot. These robots are, amongst other things, programmed to learn emotional grounded symbols. When they hear the symbol's name spoken, they know which behavior is associated with that symbol. Ronald Arkin maintains that the robots in a sense know what the symbol stands for, "in a way in which it affects both [their] internal state and what behaviors are the correct ones to use in the associated object's presence" (Arkin 2005). This implies, of course, that these androids recognize speech which allows them to behave emotionally. Charles Rubin maintained that computers will "at the very least, be capable of responding to stimuli in ways that are indistinguishable from human responses" (Rubin 2003).

Today, androids and humanoid robots can perform tasks that are typically human. For example they may be used as personal assistants, helping the sick and elderly. They are 'hired' to do the dirty or dangerous jobs. Also some administrative jobs such as being a receptionist are suitable for androids. The goal of robotic design is to have them perform any task a human being can do.

However, they are not perfect yet, but they are nevertheless already very similar to human beings. They exercise eye-movements, similar to humans. For instance, in an experiment on human-android interaction, the android did not look into the human's eyes when telling a lie, but as humans do, eyed-away (Ishiguro 2005: 5). They use their faces to express different emotions, such as, e.g., being sad or happy, being angry and wondering. They breathe (chest movements due to air actuators) and move around in and perceive their environment similar to humans. Due to neural networks they learn and develop more and more skills. In other words, similar to human DNA which is seen as the basic program for human development, the robots have a basic program allowing them to learn and develop. They are that similar to human beings that, 70 % of the subjects of an experiment on human-robot interaction did not recognize the robot as a robot immediately. Indeed, recognizing the robot as a robot took several minutes (Ishiguro 2005: 3).

Hiroshi Ishiguro, one of the most prominent android scientists in Japan, concluded his study on human-robot interaction with the words: "If human unconsciously recognizes the android as a human, he/she will deal with it as a social partner even if he/she consciously recognizes it as a robot" (Ishiguro 2005: 5). He argues that, having a growing elderly population, Japanese scientists hope to be able to fully develop androids in order to have them applied as assistants in elderly care and medical environments.

Let us now return to Searle and the question of intelligent machines. Are androids intelligent? I think we can answer this question in the positive. One reason is that androids do not manipulate data as AI machines of the 80s and 90s did. Since they start their robot life with a basic DNA-like program enabling them to learn from human-robot and robot-environment interactions they become more and more skilled and developed. They are able to act as intelligent agents in a human manner. However, that does not imply that there is no difference between human intelligence and android intelligence, especially when it comes to emotional intelligence (EI). In order to obtain EI, androids need to have a kind of theory of mind. They may recognize emotional symbols and may react to these symbols in an adequate manner; they do not do so on the basis of 'you are like me'. Furthermore, even if an android may show compassion, that does not mean that the robot feels compassionate. If a robot shows a sad face, that does not mean that the robot is feeling sad. Also imagine a robot down on its knees and praying. To whom would the robot pray? A robot may seem to be in love, behave like a human being in love but does the robot feel butterflies in its stomach or think about the other person all the time? As John Puddefoot puts it,

> To be regarded as something approaching the human, a computer [android] would need to grow, feel pain, experience and react to finitude, and generally enter into the same state of mixed joy and sorrow as a human being. In particular it would need to be finite, aware of its finitude, and condemned one day to die (Puddefoot 1996: 92).

Hence, the philosophical question should no longer be 'can a machine think', rather it should become 'can a machine feel'.

Conclusion

Even if androids possess intelligence, i.e. they can think in an artificial sense, and even if androids possess social abilities in the sense that they interact with humans and their environments, they still do not possess emotional intelligence and it is difficult to see how they would obtain such. The emotions occurring between the

Android – and human being

	Android being	Human being
Ens	Body	Body
	Neural network	Neural network
	Initial program, including intelligence	Initial program (DNA), innate intelligence
	Body and facial movements	Body and facial movements
	Eye movements	Eye movements
	Breathing	Breathing
	Heart beat	Heart beat
	Audio and visual capacities	Audio and visual capacities
Esse	Environment recognition	Environment recognition but also knowing one's place in it
	Speech (singing)	Speech (singing)
	External emotional and affective expressions (Symbol recognizing and acting upon correctly)	External and internal emotional and affective expressions Theory of mind
	External intentionality	Internal intentionality
		Introspection
		Experiencing

android and the human derive from the human only. While a human ens can be copied, human esse cannot be entirely copied, at least not yet. Androids are entirely externally oriented while humans are both externally and internally oriented. Humans have introspection; they consciously experience. They can share their experiences something that is not (yet) possible for androids. I want to close this chapter by repeating Green,

> Man does not have a soma, he is soma. Indeed, man, his person as a whole, can be denoted by soma. Man is called soma in respect to his being able to make himself the object of his own action or experience himself as the subject to whom something happens. He can be called soma, that is, as having a relationship to himself – a being able in a certain sense to distinguish himself from himself. The human person does not consist of two (or three) parts, then, but is a living whole (Green 2008: chap. 1).

Nevertheless, I just wonder though, what would happen if computer scientists discover the possibility of replicating the theory of mind?

Conclusion Part I

The first part of this book concerned what human being is about. Obviously what is meant by the thorny concept 'self' needed to be considered. Also the question of what we mean by human being needed to be answered. In order to answer this question, I investigated whether and in what sense human beings are unique, and what would be the sine qua non of human being.

I argued for the view that the mind and the brain are two sides of the same coin called human being, that there is mutual causation that influences each side of the coin in a real sense, that there is some kind of overlap between the mental and the physical. I suggested two elements of being, namely ens (a being) and esse (being), because, human beings are not only advanced neocortices and DNA (ens), they are mothers and fathers, colleagues and friends; they love and hate, and strive for better life-conditions. They paint, write poetry, create music, build monuments and advanced houses, build societies and institutions. And many are religious. They also are 'esse'.

Humans experience having some kind of a self, some kind of an 'I' that 'feels' like a deeper or elevated part of the mind and body. I argued that the self is a complex phenomenon that refers to our experiences, our relations with and understanding of us, the world and, for many people, God or ultimate reality, our feelings and emotions, our attitudes and behavior, our thinking, dreaming, and our uniqueness. Therefore we expect the underlying and correlating neural activity to be complex as well, if it is to sustain the self in all its expressions. This was confirmed by neuroscientific studies of the self in terms of complexity, embodied within the entire hierarchical system across multiple hierarchical neural levels. I suggested that what we call the self is a phenomenon of strong emergence. I argued that there is but one process which being of a specific complexity has the quality of a self. I argued that mental phenomena do play a causal role (e.g. piano players enhancing their brain). I also argued for the view that mental states, similar to physical states, have their legitimate place in space and time; it is real and is itself a multiple level hierarchical property.

By looking at neuroscientific studies I suggested an *emergent self (ES)*, containing the objective neural self (ONS) and the subjective neural self (SNS) between which there is mutual causation. ES(ONS $\leftrightarrow$ SNS). By looking at clinical and experimental neuroscientific studies I could establish a third parameter of the emergent self which I called *the Subjective Transcendent Self* (STS). The three parameters of the self being in place, I postulated a (strong) emergent three-fold self (ES) that comprises an Objective Neural Self (ONS), a Subjective Neural Self (SNS) and a Subjective Transcendent Self (STS). The relationship between the three elements is as follows:

$$\mathrm{ES}\left(\mathrm{ONS} \leftrightarrow \left(\mathrm{SNS} \cap \mathrm{STS}\right)\right);\mathrm{STS} > \left(\mathrm{ONS} \cap \mathrm{SNS}\right).$$

The function of the ONS is to neurologically sustain the subjective selves. The function of the SNS then is to express the neural self. Finally, the task of the STS is to be the essential observing subjective self, transcending the former two. The STS then is seen as the part of the self that always was and always is, itself, irreducible to neither the neural self nor the subjective neural self. By way of mutual causation, the three elements of the self trigger an emergent process of the whole self (ES).

In search for the sine qua non of human being (esse) I suggested that the sine qua non of human being has to be to experience. Human beings cannot not experience.

By human experiences I intended all experiences a human being is able to have. Since a human being is related to the world, I designated a human experience as an experience of things and events whose existence ultimately is constituted of properties and relations, actions and interactions of whatever the physical and psychological worlds treat of. Experiences are what constitute our reality. Experiences are always absolute, they are unique and individual. Experiences are real, something which I tried to show by making a distinction between the terms concept, conception and conceiving. That experiences are real does not mean that the reality of our experiences, i.e. how the world appears to us, is how the world really is. The reality of the experiences, so it seems, is not (always) as we conceive of it. Experiences have consequences, whatever behavior comes from whatever experience, it will bring about new experiences for us and others, shaping new behaviors shaping new experiences. Human beings and other animals also, cannot exist without experiences. One difference between human beings and animals is that humans have the ability to 'analyze' their experiences and understand why they experience. They understand that others also experience and compare their experiences with others. Obviously, this raises ontological and epistemological questions.

I argued that all experiences are subjectively ontologically real. Experiences cannot always be epistemologically justified but can be non-epistemologically justified in the sense of being ex-past justified. I argued that the human species differs from other animal species because of its advanced neocortex and differences in DNA compared to other species of the animal kingdom.

Concerning the uniqueness of human esse, I compared human and non-human sociology. The criteria for being a social animal in the true sense I suggested to be *firstly*, to possess a self-identity as well as a group-identity and a self-identity within a group. *Secondly*, to understand intentionality as well as understanding collective intentionality something which requires possessing the Theory of Mind. *Thirdly*, the members of a group have to be committed to the joint action. I suggested three types of animals that possess self- and group identity. The first group consists of animals that are part of a social group but are part of it as individuals. They may have their specific places within the group. It was suggested that these animals have a sense of 'I' that has a specific task within a we. These animals were said to be *more self-centered than group-centered.* Most mammals belong to this group. The second group consists of animals having a group-identity but no or little self-identity. These animals are *group-centered rather than self-centered.* Ants belong to this group of animals. The third group consists of animals that have *both self- and group identity*. They are able to think in an *I-we-mode* as well as in a *we-mode that includes the I-mode*. Some mammals (the great apes and dolphins) and humans belong to this third group. Concerning intentionality, it was shown that there is at least some evidence that some non-human animals possess self and group identity and that they may understand intentionality. In order to possess collective intentionality, it was argued that the following criteria are necessary, mutual responsiveness, a form of coordination between the members, sensitivity to each other's behavior, understanding of each other's action as a specific intentional action which one is able to respond to, and commitment to the joint action. Several experiments showed that some non-human animals fulfill the criteria for possessing self and group identity, intentionality and collective intentionality. Still, I argued that there is something unique to human social being. Humans are not only social animals; they are also cultural or institutional animals. Humans not only live in societies, they create their societies; they create culture and institutions. Only human social ontology includes deontic powers, which have their function not on the basis of their physical appearance, but on the basis of a collective acceptance. Human beings are creators.

One creation I investigated was the creation of androids and how 'human' they are or could become. I proposed that even though androids possess intelligence and social abilities, they do not possess emotional intelligence. While a human ens can be copied, human esse cannot entirely be copied, at least not yet. Androids are entirely externally oriented while humans are both externally and internally oriented.

Part II
The World

To see a world in a grain of sand,
And a heaven in a wild flower,
Hold infinity in the palm of your hand,
And eternity in an hour.

William Blake

Introduction

The aim of the first part of the present manuscript concerned human being. The second then continues this discussion but focuses on how human beings relate to the world. In order to start the enquiry on human being, it was necessary to look at what more exactly is meant by 'being'. Similarly, in order to start the enquiry on human being in the world, it is necessary to look at how reality is understood, something which is the topic of the first chapter below. The second and third chapter investigates human behavior towards others, i.e. empathic and compassionate behavior but also bad or even evil behavior. Since human beings live in space and time, the last chapter will tackle the problem of time.

Chapter 4
Understanding Reality

Introduction

What can we say about reality as such, the reality of the world around us? How can we explain, understand our universe and everything in it? As argued in chapter two, human experiences are real but real on different levels. Some are merely real in a subjective way while others real in an intersubjective manner. Because humans are intertwined with the reality they live in, they cannot describe reality as it really is. They have to conceptualize reality in order to understand and describe it. It is here the philosophical question comes in, what do we mean by the concept of reality?

The question of what we mean by the concept of reality has taken up the attention of philosophers for many centuries. As a matter of fact the question, how we can or should understand reality is one of the oldest questions philosophers have asked. Hence, perhaps it is not so unexpected that we have ended up with a variety of opinions of reality which are discussed and described within the philosophical debate called realism. Within this debate philosophers defend *metaphysical or naïve realism, critical realism, anti-realism, non-realism, internal realism, pragmatic realism, non-metaphysical realism, scientific (including materialistic) realism, semantic realism, epistemic realism, theological realism and moral realism.*

However, perhaps one could argue that there are two types of realism that are especially debated, not least amongst philosophers of religion, namely, metaphysical or naïve realism and anti-realism. Metaphysical realism says that the world is exactly as we conceive of it. Anti-realism wants us to believe that there is no objective reality. In other words they argue that there is no fact of the matter as to whether or not P. Surely, these understandings are not satisfactory.

The question is what, more exactly, is missing in these views?

A.L.C. Runehov, *The Human Being, the World and God*,
DOI 10.1007/978-3-319-44392-8_4

Reality and Realism

That reality is a tricky matter becomes clear if we know, as mentioned earlier, that scientists are in agreement that only 4 % of the universe is known. The remaining unknown 96 % is said to consist of some unknown *dark matter or energy,* some kind of *subatomic particles* (Overbye 1991). In other words, only four percent of the universe comes into our reality, 96 % does not. As if this was not limited enough for how human beings are able to talk about reality, it seems that a huge part of the human brain does not reach human consciousness. As also mentioned in the first part, the human brain seems to have some *dark energy,* called a *default mode network (DMN)* (Raichle 2010, 28–33). To complicate reality even more, and as I have also mentioned above, colors are real from the perspective of human perception, but according to physicists, colors as such do not *exist* (Ward 2005, 207). The presuppositions for humans to experience colors are in place, but there is no red, blue, yellow color like humans experience it. The same counts for gravity and time. This means that how we perceive the world is (perhaps always or at least often) different from the way the world really is. Hence, how we conceive of the world has an impact on *our* opinion of reality. Let us now return to the types of realism and the question why metaphysical or naïve realism and anti-realism are not fruitful understandings of reality?

Metaphysical – and Anti-realism – Internal Realism

Metaphysical Realism

Those who defend metaphysical or naïve realism argue that there actually is one reality that is independent of whether or not we observe it. I do not have a problem in agreeing with this statement because, if there was no reality independent of whether we observe it or not, we would not be able to observe any reality at all. After all, 4 % of the universe has been discovered (observed) and 96 % is open to discovery. Also, the fact that we need to give everything a name, i.e. that we need to conceptualize reality, is not the same as to say that reality is equal to our conceptualizations. In other words, there is more to reality than what we have conceptualized. My concern lies with the metaphysical realists' second argument, that there is one true description of this reality. They mean that one tells the truth when one's statements correspond one to one with the conditions of this independent reality.

This might be true for artifacts, like e.g. the wooden table in front of me, or the book on my desk, and also for other species, e.g. cats, or fellow human beings. If I say that the wooden table in front of me is made of oak, it is one and a half meters long and ninety centimeters wide, and has four metal legs, I have described this table in a way that it corresponds one to one with the conditions of the table. Even if human beings conceptualized this artifact as a table, they can be pretty much certain that what they describe, is as they describe it. Humans are not part of that type of reality; they are not part of the tree, the cat, the wooden table.

It becomes different when we try to describe the reality we are part of, chemical – physical – quantum realities. In this case metaphysical realism cannot be applied, because that would mean that human beings have a capacity to see reality from *nowhere* or *with God's eyes.*[1] In other words, human beings would possess the capacity to raise themselves above a reality they are part of and to describe reality from this *outer perspective.* There are at least two problems with endorsing such a type of realism. Firstly, metaphysical realism leads to skepticism and secondly it is pragmatically seen as incorrect. (Herrmann 2008). The second objection that metaphysical realism is pragmatically not tenable is relatively easy to support.

We do not need to refer to quantum physics in order to see that metaphysical realism cannot correctly describe a reality in which we are intertwined. I have already mentioned the phenomena of colors and gravity, but there is also the phenomenon of time. However, looking at quantum physics, the fact that there are different interpretations of the same reality suggests the statement, that there can only be one true description of reality, cannot be correct.[2]

For instance, Hans Bohr quoted his father Nobel Prize winner in physics 1922, Niels Bohr (1885–1962) as follows, "[…] profound truths [are] recognized by the fact that the opposite is also a profound truth, […]" (Bohr 1967, 328).

Niels Bohr's own interpretation, the *Copenhagen interpretation,* suggests that reality, which in the language of quantum physics is called the 'superposition (Ψ)' collapses at the moment of measurement or observation.[3] Ψ is interpreted in terms of *probability*, which means for example that we can only specify a probable location of the particle. Hence, from the perspective of quantum physics it is difficult to defend the possibility of being able to employ a 'God's-eye perspective' of reality.

---

[1] This is not to be confused with Thomas Nagel's view of realism. He argues that we cannot grasp the world as it is due to limited knowledge as well as to what we are able to conceive. In his words: "The world extends beyond the reach of our minds" (Nagel 1986: 90).

[2] There are several interpretations of the Quantum Mechanical Theory: the micro-macro interpretation along the Bohr line, better known as the Copenhagen Interpretation, the mind-body interpretation along the Wigner line also known as the Princeton interpretation, The Many Worlds interpretation of Everett III, the Many Minds interpretation by Albert & Loewer and later David Deutsch. Furthermore there is the interpretation by Penrose taking gravitation as the cause of the collapse, and there is the Pilote Wave interpretation along the de Broglie-Bohm line. Furthermore, there are the more recent interpretations, for example the modal interpretation and the interpretation based on decoherence.

[3] The basic physical quantity used in quantum mechanics is the wave function, i.e. Ψ as described by the Schrödinger equation. The Schrödinger equation tells us that, if we know Ψ at time zero and we know the Hamiltonian, then we can calculate Ψ for all future times. The first postulate of quantum mechanics is that all information about a physical system is to be found in the wave function, i.e. Ψ of that system. If we manipulate the wave function Ψ as we measure and/or observe a particle, then we receive information about the location, energy, velocity of that particle. It might be interesting to note that Schrödinger gave the wave function Ψ an ontological meaning; he meant that the wave function Ψ exists independently of our knowledge about it. Also von Neumann had a realistic view on the wave function. Bohr and Heisenberg on the other hand had a non-realistic view on the wave function and hence avoided the measurement problem.

But this not only applies to quantum physics. When it comes to explaining reality, scientists need to look at a limited part of reality and explain this limited part in terms of mathematical, biological or chemical formulas without arguing that there is a true relationship between all of reality and their descriptions. Does this mean that 'everything goes', that reality is relative to human conception and descriptions? Does this mean that anti-realism is a more fruitful approach to reality after all?

Anti-realism

Those defending anti-realism also known as non-realism argue that all of reality is merely constructed by us. Don Cupitt, who is perhaps the best known representative of anti-realism, argues that we cannot have any knowledge about reality at all. We can only have knowledge about our own reality which is built upon our ideas, viewed from our own perspective and constructed to fill our personal needs. He even goes so far as to maintain that we cannot have true knowledge about ourselves. In *Is nothing Sacred,* Cupitt writes that there simply cannot be another view of the world than the view we have systematically developed ourselves (2002: 51). He continues, "What keeps truth true is the vividness with which it is re-imagined and re-expressed" (2002: 52).

Once again, there is some truth in this view, for example when it comes to subjective opinions of reality, like e.g. Shirley's pet arm. But that does not imply that all our perceptions of reality are purely subjective, or that these are only real for us. After all, humans share their experiences of the world and for many (though not all) experiences there is consensus amongst the experiencers about what they experience. Having an adequate set-up of conceptions, these can be communicated.

Hence, neither metaphysical nor anti-realism seem to be fruitful approaches in explaining or understanding reality. The question is; what type of realism would best describe the relationship between what is out there and what humans perceive and describe?

Take stars for instance. A star is a luminous sphere of plasma held together by its own gravity. It starts with a gravitational collapse of a gaseous nebula of material composed primarily of hydrogen, helium and trace amounts of heavier elements. After x-amount of time, everything comes into place and the star releases energy. Seen from Earth, the closest star is the sun. Seen as such (physical reality) stars are a bunch of chemistry. But that is not how humans always 'conceive' of stars. Think about the star of Bethlehem, a falling star, the morning and evening star and further how 'we wish upon a star', how we compliment someone, saying 'you are a star', etc. This means that besides the chemical/physical component of the stars, people tend to give stars a phenomenological component in the sense of what stars *mean for them.*

This autumn we had the opportunity to experience the phenomenon called a blood moon, also called a super-moon lunar eclipse (Sept. 27–28). This natural phenomenon rarely happens. It seems we will have to wait until 2033 to see another one. This natural phenomenon happens when the earth passes between the sun and the moon. Then earth casts its shadow far out into space, beyond the orbit of the moon. The earth's shadow is red at the edges for the same reason a sunset is red, i.e. the red color is caused by Rayleigh scattering of sunlight through the Earth's atmosphere. What we observe is a big round red/orange moon, as if the moon has come much closer to us. The view is amazing, of course. Furthermore, similar to the Bethlehem star, the red moon has a religious meaning. For some it means the coming of the end of times, as it is written in the book of Joel, "the sun will turn into darkness, and the moon into blood, before the great and terrible day of the Lord comes" (Joel, 2–31). Others put a romantic dimension to the phenomenon.

Internal Realism

In order to tackle this problem of reality, some philosophers adapt internal or pragmatic realism, meaning that what is real, is what is *real for us*, without neglecting that there is more to say, but it is not always fruitful to try to say something about that. Philosophers who thought along these lines were Immanuel Kant, (1724–1804) who argued that we cannot have knowledge about Das Ding am sig ([1781] 2004). The early Ludwig Wittgenstein (1889–1951) argued that what we cannot know something about we should not speak about either ([1921], 1992). The early Hillary Putnam claimed that although the world may be causally independent of the human mind, the structure of the world, its division into kinds, individuals and categories is a function of the human mind, and hence the world is not ontologically independent. (1998). For instance, Putnam was inspired by Kant. A scientist who might be interpreted as having an internal realistic view of reality is Bohr, when he says that nothing exists unless observed or measured (I will return to this in the chapter on time). Neither did he mean that there does not exist more than what we observe, only that we can only describe what we observe.

The question is raised whether internal or pragmatic realism is the most fruitful type of realism in order to describe reality? My answer would be yes and not really, because in my opinion, human reality is far more complicated than any of the types of realism embraced.

Hence, I suggest yet another type of realism which I call *extended realism* which includes three dimensions to how human beings perceive reality: (1) a measurable or observable reality; (2) a creational reality and (3) a phenomenological reality.

Extended Realism

Measurable or Observable Reality or Reality for Us

This dimension includes e.g. black holes, rain, DNA, stemcells, etc. This dimension is *intersubjective*; in other words, it is described by all or at least a majority in the same or similar way. This dimension concerns the physical, biological, neurological, chemical world etc. we explore. While I argue that this dimension is intersubjective, some (but not all) scientists maintain that what they refer to is objective in scope. Also psychologists whose aim is to understand the subjective quality of life of a person still argue that they study subjective qualities in an objective, quantifiable manner, similar to e.g. biologists studying bacteria.

In a previous article, psychologist Harald Walach and I considered three main approaches concerning psychological observations and its reality (2007: 65). On the one hand we have the auto-phenomenological approach, also known as naïve objectivism, as defended by Charles T. Tart. In this approach introspective experiences are explored (and explained) from a first-person perspective (2006). On the other hand we have the deconstructivist subjectivity approach which denies an essential or intrinsic meaning of the experience. We could say that William A. Adams is in favor of such an approach (2006).[4]

In between we find the hetero-phenomenological approach as defended by Anthony Freeman. The idea is to explore subjective experiences from a third-person perspective (2006). In our opinion, neither approach works, because the auto-phenomenological approach is limited in that it is loaded with paradoxes and reflexivity (Oviedo 2006: 391), while the hetero-phenomenological approach is limited because it cannot escape characteristics of autophenomenology.

Furthermore both the scientist and the observable are, as it were, in the same position, namely, within the same universe. Hence, there is always some kind of relationship between the subject of study and the researcher. For example, the bacterium 'observes' the researcher in its specific biological manner and the scientist observes the bacterium in his or her specific biological manner. The scientist, be it a biologist or psychologist, is bound to his or her senses to describe reality.

This becomes most clear within the discipline of quantum physics. Indeed the measurements performed by quantum physicists cannot be separated from the measurer. Quantum mechanically, this is interpreted as it is projected onto the observer's senses. The fact that vast numbers of physical observers experience these projections in a similar way is interpreted as depending on two circumstances: (a) Besides the observers themselves, the outer universe is the same for all observers (assumption), and (b) the universe is comprehended similarly by all observers due to them having a similar biological construction (observation) (Holte, R., 1984: 56–57). In other words, this dimension of reality is not free from the observer's senses and can therefore not be an objective dimension but an intersubjective dimension of reality.

Let us return to the example of gravitation. We all know gravitation from our experience. We have a good physical theory that describes gravitation. This theory also predicts an exchange particle for gravitation as a force, the graviton. This

[4]Deconstructivism derives from the philosophy of language of Jacques Derrida (1982). A word receives it meaning as a function of how it differs from another word. What Derrida means is that complete meaning is always differential and postponed in language; there is never a moment when meaning is complete and total (Glendinning 2004: 7). Richard Rorty (1931–2007) argued that an "'accidental' feature of a text can be seen as betraying, subverting, it's purportedly 'essential message'" (1995). Translated to deconstructivist subjectivity of experiences this means that the reports of the experiences have little to do with reality.

particle, however, has not been found so far. In other words, the graviton has a hypothetical existence. Gravitation is *real* in the sense that all people on earth experience it as real. For astronauts who experience the loss of gravitation while in space, due to the loss of the force of gravitation, non-gravitation is real. Hence, the reality of the graviton is not separable from the theoretical and biological structure that we use to experience and understand the world we are living in. We assume that we live in a reality that interacts with our senses in a way similar for everyone (we experience gravitation on Earth; we do not experience it in space). Hence, we assume that all observers experience gravitation in the same way because we rely on the observation that all observers have a similar biological construction. These are the starting points from which theories are built and further empirically tested.

This also means that a distinction needs to be made between what is predicted (hypothesized) as a theoretical entity (for example, a graviton), and a theory or the whole process of negotiation.

1. All people on Earth witness gravitation in the same or similar way.
2. Scientific experiments confirm the material basis for gravitation.
3. Consequently, gravitation exists.

However, that all people on Earth witness gravitation in the same or similar way (premise 1) does not imply that gravitation *exists,* nor does the material (scientific) basis of gravitation (premise 2) prove anything of the like, because what this material basis does prove is only that there seems to be a specific relationship between organisms and gravitation. Astronauts do not experience gravitation when out in space. The planet Mars does not have the same degree of gravity as planet Earth. On the other hand, astronauts landing on Mars are the same astronauts as they were before take-off (except perhaps for personal individual changes, biological ones due to not having been able to eat properly, and other changes due to being excited about the whole enterprise). The tools they bring with them are the same on Mars as on Earth (even though they might behave differently). In this sense, one might argue that the astronauts and their tools exist even when gravitation does not. Astronauts cannot bring gravitation or a graviton with them while traveling in space.

What we can say is that the experience of gravitation is real through its intersubjective verification status and through its material basis discovered by scientists. It could be argued that all scientific statements about reality that eventually lead to something which is scientifically accepted are subjective accounts to begin with. Fleck looking through his microscope, Galileo Galilei (1564–1642) looking through his telescope, Werner Heisenberg (1901–1976) or Erwin Schrödinger (1887–1961) observing the movement of particles, the participant in a cognition experiment recounting the syllables he or she remembers, they all make subjective statements. The first thing that is reported is the researcher's own experience or intuition. Once on the scientific table then, these experiences and intuitions are verified.

Verification is one of the main criteria for science.[5] Another important criterion is repeatability, which implies that experiments, observations or measurements can be repeated by other scientists under the same or very similar conditions. Once these steps are confirmed the scientific statements of reality become intersubjectively confirmed.

However, the fact that the reality described by scientists is based on intersubjective observation does mean that there is no objectivity at all involved. To a certain extent there is in the sense that this dimension of reality lives its own life following the laws of physics (or biology or neurology etc.).

Creational Reality or Reality from Us for Us

Human beings have the capacity to create their own reality from the reality they are part of. I am not only thinking about all the artifacts created by humans; rather I am thinking of the social reality humans created.

To repeat what is said above, this is a dimension of reality that has its function not on the basis of its physical appearance, but on the basis of a collective acceptance. For example, the function does not lie in the paper sheet of a $100 note, but in its value of $100. In other words, the function lies in the collective agreement that this paper sheet is worth $100. Other examples of collectives agreed upon reality are contracts, professions, wedding-rings, corporations, the stock market, religion, etc. As mentioned in part II, John Searle argued that this dimension of reality has specific 'deontic powers' (Searle 2008: 33).

This dimension of reality is completely conceptualized by us. This dimension of reality is a reality from us for us. Nevertheless, it also lives its own life and seems to follow the laws of nature.

It is believed that the world's first stock exchange was established in Bruges (Belgium) 1309. The term developed to become a sophisticated currency market. But we had to wait until the 1930s before the concept of economic crash was coined. The market became overheated and exploded, more or less like an eruption. From this experience, safer control systems were created etc., which means that the reality humans created became measurable or observable and hence provided resistance to its creators. Today, we know that nothing much needs to happen in order for the global currency market to shudder. We could compare this to a butterfly somewhere in Australia triggering an eruption in Iceland. Perhaps chaos theory may also be

[5] Some still seem to hold on to a Popperian view of science which stipulates that only falsifiability is a viable methodological path. While Popper's arguments have long since been refuted, for instance by Putnam and others, they are tenacious in some quarters, because they have teamed up in an unhealthy alliance with Fisherian statistics that allow only for deductive types of hypothesis testing which in turn look similar to falsifiability. Rarely is there a critical distinction made between the theory of science type of claim Karl Popper (1902–1994) makes, and the statistical procedure adopted, let alone their relative merits critically scrutinized. Hence we still hold on to the seemingly old-fashioned view that science has to do with verification, although certainly not in its original, simplistic and naïve positivist framework. See further Popper, K., 2005.

applied to the human created dimension. The created dimension of reality develops fast.

This dimension of reality includes religious and political systems, institutions, different authorities, systems of law and so forth. All may lead to a good life, but may also destroy our lives.

Neither the measurable nor the created dimension of reality would mean anything to us if it were not for a third dimension, namely, the phenomenological dimension of reality.

Phenomenological Reality or Reality Within Us for Us

We find this dimension of reality within us. Without it, we would not be able to conceptualize reality or to create reality. It is a dimension of reality that is individual and dependent on every person's neurological proposition. Hence, we could say that this dimension of reality follows neurological laws. Nonetheless, it is conceptualized by us, which is necessary for communication. In this sense it is subjective as well as intersubjective. This dimension of reality is the reality of our experiences, (within us) for us.[6]

In conclusion, human beings have to cope with different dimensions of reality; they cannot withdraw themselves from these realities. All dimensions are real for human beings in different ways and to different degrees.

The Relationship Dimensions of Reality

I suggested three dimensions of reality:

1. Measurable/observable reality or reality for us (MR). E.g. black holes, rain, DNA, tsunami, stemcells. This dimension of reality is intersubjective. This dimension of reality corresponds to our measurements and observations, i.e. our conceptualizations.
2. Created reality or reality from us for us (CR). E.g. currency, contracts, religion, wedding-rings. This dimension of reality is also intersubjective. This dimension of reality also corresponds to our conceptualizations because of its deontic power.
3. Phenomenological reality or reality within us (PR). E.g. love, sorrow, dreams, wishes, pain. This dimension is subjective but to some extent also intersubjective, because of our conceptualizations that make communication possible. However this dimension of reality also has a pure subjective dimension, e.g. reality caused by brain disorders, narcotics, and so forth.

What humans call reality emerges at the interface of three different kinds of experiencing of inner and outer dimensions of life (Fig. 4.1).

[6] See again chapter two on the human experiencer.

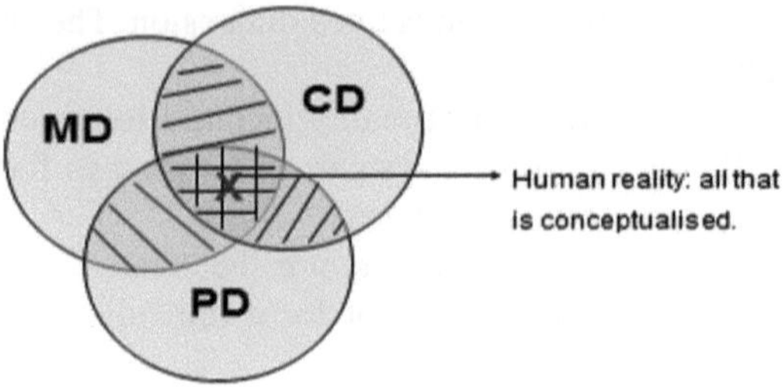

Fig. 4.1 Human reality as it is related to different aspects of reality

Naturalism

Having presented some views on reality and realism, it may be a good idea to also provide some views on naturalism. Like realism, naturalism is a widely discussed topic within philosophy. Hence, like realism, it is not possible to account for the whole debate. The term naturalism was coined 400 years ago, and denoted "a view of the world, and of [human] relation to it, in which only the operation of natural (as opposed to supernatural or spiritual) laws and forces is admitted or assumed" (Flanagan 2006: 432).[7]

In modern times, the term is associated with the twentieth century philosophers John Dewey[8] (1859–1952), Ernest Nagel[9] (1901–1985), Sidney Hook[10] (1902–1989) and Roy Wood Sellars[11] (1880–1973) who wanted to ally philosophy more closely to science (Papineau 2007). Their definition does not differ remarkably from the initial seventeenth century one. Naturalism came to denote that "Reality is *exhausted* by nature, containing nothing supernatural, [hence] the scientific method should be used to investigate *all areas of reality*, including the human spirit" (Papineau 2007).[12]

Stewart Goetz and Charles Taliaferro call this type of naturalism strict naturalism and argue that "strict naturalism, as an ideal scientific philosophy, seems to include all aspects of reality within a comprehensive and unified perspective that excludes anything that is either conscious, psychological, or mental in nature" (Goetz and Taliaferro 2008:16). David Papineau and Owen Flanagan maintain that the term naturalism has no very precise meaning or lacks a single determinate meaning (Papineau 2007; Flanagan 2006: 432–452).

Like realism, naturalism is in itself a normative position since it concerns the terms in which reality ought to be understood. From the fact that there is no consensus of how reality should be understood, it follows that we have a variety of types of

---

[7] I changed 'man's' into 'human'.

[8] *The Collected Works of John Dewey,* Jo Ann Boydston, ed., 37 volumes (Carbondale: Southern Illinois University Press, 1967–1991).

[9] Nagel 1986.

[10] Hook 1961.

[11] Sellars 1927.

[12] My italics.

naturalism to choose from. From what we shall call the pure types of naturalism, ontological, methodological and epistemological naturalism as well as supernaturalism, a wide variety of new types have emerged. I believe that the most well-known types of naturalism are ontological naturalism, methodological naturalism, epistemological naturalism and supernaturalism.

The evaluation of these types of naturalism inspired me to argue for two types of ontological naturalism that are most fruitful in order to understand reality. I coined these types, *minimalistic ontological naturalism* and *Extended* or *flexible ontological naturalism.*

This section ends with comparing two methods of understanding reality, namely, realism and naturalism, on their similarities and differences.

Ontological Naturalism

Some advocates of *ontological naturalism* argue that mental phenomena are equal to natural phenomena as described by the natural sciences. They argue that mental phenomena are physical/material phenomena, $\forall x(Px \rightarrow Qx)$, where P denotes mental states and Q all that is physical or material. Let us call this variety of naturalism, *comprehensive ontological naturalism* that implies that there is nothing over and above the physical and material.

Advocates of such a view are e.g. Patricia Churchland (2002) and Daniel Dennett (1993). Reality P can be entirely explained in terms of *nature* Q. Hence if P then Q, where Q embodies all that is matter and physical and P may embody all that is matter, physical subjective and spiritual. Hence, $\forall x(Px \rightarrow Qx)$, all P are (reducible to) Q. Obviously, there is something wrong with assuming $\forall x(Px \rightarrow Qx)$. To give an example, few, I believe, would argue that music is nothing but sound waves, i.e., air that vibrates in different sequences and with different amplitudes, because music can only occur when these sound waves meet a listener who interprets the sounds and comprehends them, according to his or her cultural, biological, and personal background, as harmonies and disharmonies (Runehov 2007: 212). Secondly, as already discussed in the section on realism, neither scientists nor scholars have any means to account for reality as a whole. They may be able to account for the particulars of reality (for example biological or historical facts), but from this it does not logically follow that they are able to account for reality as a whole. From $\exists x(Px \rightarrow Qx)$ it does not follow that $\forall x(Px \rightarrow Qx)$.

Another variety of ontological naturalism is advocated by Samuel Guttenplan, who has a more flexible view on naturalism when he says that naturalism should be seen within the frame of some *realm*. Everything that exists and takes place in that realm consists of the empirically accessible features of the world (Guttenplan 2001: 449). The realms may be the *laws* and/or *theories* of the natural sciences. Guttenplan's view on naturalism can be formulated as it is within a specific realm R that $\forall x(Px \rightarrow Qx)$. I will refer to this kind of naturalism as *minimalist ontological naturalism* (Walach and Runehov 2010: 159).[13] I will return to this type of naturalism below.

Yet other defenders of *ontological naturalism* argue that there is no distinctive metaphysical area (realm) of inquiry. There is only one natural order and it *comprises* all of reality. Nothing can be explained independently of the natural order. However, which theory is the best one to explain the natural order is not determined (Jacobs 2008). Hence, $\forall x(Px \vee Qx)$.[14] Let us call this version of ontological naturalism *extended or flexible ontological naturalism.* I will also return to this type of naturalism below.

Willem Drees, amongst others, advocates an ontological naturalistic position of the type I called extended or flexible ontological naturalism. He argues that "[t]he natural world is the whole of reality that we know of and interact with; no supernatural or spiritual realm *distinct* from the natural world shows up *within* our natural world, not even in the mental life of humans" (1998: 12; Rottschaefer 2001: 414; Griffin 2000: 66).[15] If I have understood Willem Drees correctly, he means that the mind also belongs to the natural order as a whole. Such ontological naturalism need not be reductive. Actually it allows a degree of holism and may include both religious and moral dimensions (Drees 2003: 595). In *Religion and Science in Context*, Drees goes even further and maintains that such a form of reduction.

> is not elimination; rather, it is integration of these phenomena into our multifaceted picture of the world. Such a reductionism is a form of holism as it reveals the pervasive coherence of reality (2010: 88).

Hence, this does not mean that everything there is can be explained in natural terms. There may be processes and events which the natural sciences, by way of their methods cannot depict, for example, experiencing colors. As Drees argues, science suggests a naturalistic view of reality but it has its limits. Extended or flexible ontological naturalists need not deny that an experience of colors is real. Furthermore within the realm of neuroscientific research, the question whether colors exist independently of experience or not is not an important issue. Similarly, it is not essential for neuroscientists to question whether the object of a religious experience exists independently of experience or not. What is important is to know how such experiences are possible and why they have the quality they have as described by the experiencer, i.e. affective and/or cognitive? We have established three types of ontological naturalism:

1. Comprehensive ontological naturalism stating that $\forall x(Px \rightarrow Qx)$,
2. Minimalist ontological naturalism stating that within a certain realm R, $\forall x(Px \rightarrow Qx)$,

---

[13] What I refer to as minimalist ontological naturalism should not be confused with Griffin's idea that naturalism can be understood in a minimal or maximal sense (Griffin 2000: 11–12. What I mean by minimalist naturalism is naturalism useful within a specific realm or scope of inquiry.

[14] Where 'v' stands for 'inclusive v'.

[15] Drees calls his ontological naturalism supernaturalistic religious naturalism. Furthermore, Drees defends the view of a physically closed universe. My italics.

3. Extended or flexible ontological naturalism, where $\forall x(Px \vee Qx)$.

From what has been said above as well as from the assumption that experiences, being the essence of what it is to be a human being, are real, I argue that in order to give justice to our experiences of reality, endorsing the minimalist or the extended/flexible ontological naturalistic view of reality is most fruitful.

Methodological Naturalism

We have arrived at the second type of naturalism to be analyzed on its validity, namely methodological naturalism. This type of naturalism emphasizes that the methods used in the natural sciences are the ideal methods and hence it is maintained that the methods of the social sciences and philosophy ought to be modeled on the methods of natural sciences. Methodological naturalists seek to apply scientific methods in non-scientific disciplines because of the success that the natural scientific methods have shown in the enterprise to describe the world. The assumption is that if the methods of the natural sciences were applied to the social sciences and philosophy, success would automatically follow.

Hence, methodological naturalism is not about what does or what does not exist; rather it is about applying the best possible method to understand the world. In other words, methodological naturalism is understood as a research strategy for analyzing things such as cells, in terms of their parts, such as, macromolecules, as well as for applying successful theories in one area, such as Darwinian evolution, to other areas, such as sociology and religion. However, methodological naturalism can be understood in other ways as well. Some would say that the aim of a methodological approach is to always apply the simplest explanation. For example, explaining a specific experience in terms of a correlation with a specific brain function. Observe that this is not the same as to *reduce* the specific experience to a specific neural activity. The aim is to concentrate on a specific set of parts for a scientific purpose without claiming that this is all there is to it.

Epistemological Naturalism

The view that epistemology should be naturalized derives from Willard Van Orman Quine (1960). The idea was to have an objective approach to epistemology by way of using the empirical sciences to investigate human epistemic activity (Goldman 1999: 598). This is because epistemological naturalism dictates that human knowledge is a natural phenomenon and hence should be studied in the same way as any other aspect of nature. Willard Van Orman Quine (1908–2000) tries to convince us that philosophy *is* a science, only one of a more general character (Quine 1960: 3–4). Philosophy is concerned (amongst other things) with epistemology, hence epistemology should be analyzed in a scientific manner. This has implied that a priori knowledge is denied by those defending this position. It has been argued that our beliefs must be well-formed and reliable hypotheses, i.e. formulated as precisely as possible in order to become subject to (empirical) experiential tests (Hançil and M. Ziemer 2000, in Spurway 2009: 88).

The epistemological question 'what can we know' is not unimportant, but it is a philosophical question.

Supernaturalism

The last type of naturalism to take into consideration is known as supernaturalism. Supernaturalism does not belong to the category of non-naturalism; rather supernaturalism takes an anti-naturalistic stand concerning questions about God and ultimate reality. By supernaturalism is meant "the invocation of an agent or force which somehow stands outside the familiar natural world and so whose doings cannot be understood as part of it" (Flanagan 2006: 432). Supernaturalism, in my opinion, suffers from similar objections as comprehensive naturalism. Let us recall Owen Flanagan's three components of supernaturalism:

1. There exists some kind of supernatural reality *outside* the natural world.[16]
2. Supernatural reality acts upon the natural world [from outside].
3. The basis of belief in supernatural reality and its acts is invisible, undiscoverable, and cannot be inferred by way of any epistemological methods (2006: 432).

There are obviously theologians who would defend all three premises. For instance, a consequence of accepting these premises together with the wish to find proof for the existence of God has led to Alvin Plantinga's philosophical endeavour to try to give proof for the *possibility* of ultimate reality by way of modal logics. (Plantinga 1982).

Modal logics are different from Aristotelian logic in the sense that they include, besides the logical notions of *true* and *false,* the logical notion of *possibility.* And since possibility cannot be proven true or false, it remains open. However, in contemporary theology the first premise that some kind of supernatural reality exists *outside* the natural world is no longer uncritically accepted. Rather, such a dualistic theistic view is strongly debated, at least within contemporary Christian theology.

Furthermore, the increasing inter-religious debate makes it difficult to argue for premise two as well. Thus, in order for supernaturalism to fit into the program, Flanagan's three premises would at least need to be rephrased, perhaps as follows:

1. There exists some kind of *ultimate reality* (UR) *within* the natural world but UR is not *reducible* to the natural world.
2. UR acts upon the natural world *from within.*

---

[16] My italics.

3. The basis of belief in UR and its acts remains ineffable but its marks or signs are not.

Some would argue that such a revised supernaturalism is no longer a supernaturalism. Perhaps they are right in some sense; perhaps such supernaturalism rather refers to what Drees calls supernatural religious naturalism.

Adequate Type(s) of Naturalism

The type of naturalism that is needed is one that would not reduce mental reality to material or physical reality. The most adequate type of naturalism discussed in the previous section is ontological naturalism. However, not all three varieties of the ontological naturalism states above are likely to be adequate. For the reasons given above, comprehensive ontological naturalism is not an adequate approach. On the other hand, minimalist and extended or flexible ontological naturalism are worth further investigation.

Minimalistic Ontological Naturalism and Extended or Flexible Ontological Naturalism

A minimalist ontological naturalistic view of reality dictates that within a *specific realm* R, if we have a mental state P then we have a physical/material state Q, i.e. within a *specific realm* R: $\forall x\ (Px \rightarrow Qx)$, where P denotes mental states and Q all that is physical or material. However, it is necessary to note that this does not mean that mental states Px are physical/material states Qx, which would imply a comprehensive ontological naturalism dictating that *all* mental phenomena can *exhaustively* be explained by physical or material phenomena.

This is not what is meant here. Rather, the specific realm is related to one network of realities that includes both physical and mental reality and hence can be justified by way of reference to some part(s) in that network. In other words, a minimalist ontological naturalism needs to be coherent (Walach and Runehov 2010).

Another fruitful view of reality would be an *extended or flexible ontological naturalism* that states that there is only one natural order and that this comprises all of reality x. Nothing can be explained independently of the natural order. This type of naturalism can be formulated as $\forall x(Px \vee x)$. This type of ontological naturalism seems to be more open than the former in that it takes account of the possibility of novel types of reality, mental and physical (Walach and Runehov 2010: 159).

Conclusion

I have presented two philosophical methods for how to comprehend reality, realism and naturalism. Supernaturalism and metaphysical or naïve realism are, in my opinion linked in some sense at least if we compare metaphysical realism as endorsed by some philosophers of religion. Both argue for there to be an agent or force which somehow stands outside common reality. However, while these metaphysical realists argue that there is a true relationship between ultimate reality and our descriptions, supernaturalists argue ultimate reality interacts with the physical world in remarkable and unique ways but cannot be understood as part of it. Hence, if I understood this correctly, supernaturalists would not argue for a true relationship between ultimate reality and our descriptions, since these belong to the natural world. In this sense, it could be argued that supernaturalism is also related to anti-realism, at least anti-realism as defended by philosophers of religion, because they argue that we cannot have any knowledge about ultimate reality at all.

I think it is quite obvious that internal or pragmatic realists would agree with both methodological and epistemological naturalists. In other words, they would agree that the most important question is not what or what does not exist. Rather the focus should be put on applying the best possible method to understand the world. And hence, they would agree with epistemological naturalists who argue that our beliefs must be formulated as precisely as possible in order to become subject to investigation. Nevertheless they would not necessarily agree that such an investigation needs to be empirical. However I also believe there is a relationship between internal or pragmatic realism and minimalistic ontological naturalism, since the latter view describes reality as ontological within a specific realm, without reducing mental reality to physical or material reality.

Extended realism and extended ontological naturalism are, in my opinion, the ones that come closest together. The former regards reality that is measurable, i.e. reality for us, creational reality, reality from us for us and phenomenological reality, reality within us for us. Extended ontological naturalism dictates that there is only one natural order comprising ALL of reality.

Chapter 5
Mindreading

Introduction

While the previous chapter concerned reality as a whole and how we can make sense of it all using different philosophical methods, the present chapter regards a specific reality within us. It concerns mindreading, i.e. what we mentally do when we interact with other people.

Human beings, it is argued, are mindreaders. Sometimes human beings are referred to as Machiavellian mindreaders. In other words, they read other minds in order to predict the behavior of others in order to outsmart them. Machiavelli argued that "for a prince […] it is not necessary to have all the [Virtuous] qualities, but it is very necessary to appear to have them" (Machiavelli 1998; Whiten and Byrne 1997: 13).[1] I believe there is little doubt such may be the case for some humans and in some situations.

More important to emphasize though is that the capacity to read other minds has given human beings various evolutionary advantages. The core of the mindreading capacity is said to be empathy. Empathy comprises, amongst other capacities, compassion. This chapter hence accounts for mindreading, empathy and compassion.

First- and Third-Person Mindreading

Mindreading is the capacity humans have to understand other minds. However, it also concerns human knowledge about the content of their own conscious mind. Human beings are able to like or dislike this content and change it accordingly. In

[1] Whiten and Byrne emphasize that Machiavellian intelligence is prominent in primates. Niccoló Machiavelli (1469–1527). See also, Premack 1988.

A.L.C. Runehov, *The Human Being, the World and God*,
DOI 10.1007/978-3-319-44392-8_5

other words, they do have first-person authority over their minds. To put it differently, humans can *read and write* their own mind (Runehov 2008a).

Hence, from a first person perspective and under normal neurological conditions, humans are not only mindreaders but also what I refer to as *mindwriters*. This means that first-person mindreading is directly given and non-inferential but vulnerable to first person modification. In relation to other people, human beings have the ability to disguise or fake their own states of mind. In the words of philosopher Pierre Jacob, relying on the classical view of mindreading held by philosophers as well as psychologists, a view I will defend, "[a] first-person mindreading consists in forming a belief about (or in metarepresenting) a psychological state of one's own" (Jacob 2009: 572). Consequently, a first-person mindwriting consists in *reflecting* about and *adapting* and/or *changing a belief* about (or in metarepresenting) a psychological state of one's own.

Third-person mindreading is different. Following the classical view of mindreading again, "a task of third-person mindreading consists in forming a belief (or judgment) about (or in metarepresenting) another's psychological state" (Jacob 2009: 572). Obviously, the task of mind-writing the other person's mind is excluded, the mind-writing task remains on a first-person level, i.e., the reader reflects about, adapts and changes their own belief or judgment of the other person's psychological state.

Let me illustrate this line of thought. Suppose I come home one evening and see my husband sitting at the table, his hands folded under his chin while gazing through the window. Being married to him for more than 20 years, I will probably form the belief that he is trying to solve a computational problem, something which is often the case when he sits like this. However, I remember that on some occasions, when he has exhibited this body-posture, he was trying to cope with his irritation about something. Hence, I may change (or may have to change) my primary belief. I may have to re-write my mind about his psychological state. What I cannot do is re-write his mind; he is in a particular psychological state and nothing else.

A necessary condition for both first- and third-person mindreading is that the "mindreader possesses and applies to the task the *concept* of the relevant psychological state (e.g., intention, belief, desire, and so on)" (Jacob 2009: 572).

Empathy

Empathy as the core subject matter of mindreading is something that is established between (human) beings. Empathy, it is argued, "is dependent on neural circuits that have evolved in the context of parental care and living in social groups" (Decety 2014: 127). Empathy belongs to third-person mindreading and is as such a mental state.[2] The first-person perspective is only important in relation to third-person mindreading.

According to Alvin Goldman we can divide mindreading into *low-level mindreading* that targets primitive mental states such as disgust and pain and is involuntary, and *high-level mindreading* that targets more complex mental states such as propositional attitudes, which is to some extent voluntary and is to some degree linked to consciousness (Goldman 2006: 113–191; Runehov and Anne 2012b).

Based on his neuroscientific studies, Rüdiger Seitz also argues that there seem to be two different orders of empathy, a more complex, cognitive level and a more automatic, imitative level of empathy (Seitz et al. 2008: 714). As mentioned above, and inspired by Goldman and Seitz, I divide empathy into low-level empathy as being automatic and imitative and high-level empathy being more complex and cognitive and linked to conscious reflection (Runehov and Anne 2012b).

Empathy further comprises (a) the capacity to affectively respond to another person; (b) the cognitive capacity to put oneself mentally in the other person's mental condition and (c) the capacity to regulate one's emotions in relation to the other person (Decety and Jackson 2007: 71–100). Philosophically, these capacities are known as the capacity of the *Theory of Mind (ToM)*.

I mentioned the ToM in chapter one, but as a reminder and briefly, this theory refers to the capacity to attribute mental states; i.e. beliefs, intents, desires, pretending, knowledge, etc., to oneself and others and to understand that others may have beliefs, desires and intentions that are different from one's own. Philosophically, we attribute mental states to ourselves and others by way of analogical inference. Roughly explained, the analogical inference principle is the idea that other human beings are 'very like me'. In the language of logic, we induce the other from ourselves. Simply put, x observes y, y is like x, hence x understands y. This is Andrew Meltzoff's theory of imitation in a nutshell (2005: 56).

I chose to divide low- and high-level empathy as follows. Low-level empathic behavior comprises emotional contagion,[3] biological altruism[4]; this is more or less primitive and/or involuntary. High-level empathic behavior comprises sympathy, selfless altruism,[5] and agape[6]; these are more developed and complex forms of

[2] Generally, a mental state is understood as that which is not in itself material, physical or neural. However, the concept of mind is problematic. Mind is understood as awareness, consciousness (including higher-consciousness, sub-consciousness and unconsciousness), the psychical and sometimes also as the spiritual. It is related to sensation, perception, reasoning, intention, belief, memory, etc.

[3] This is the capacity to mimic the facial and vocal expressions, postures and instrumental behaviors of those around us and thereby to catch the emotions of others.

[4] Is evolutionarily innate and is shared by humans as well as other species. It is involuntary and concerns the act of offering oneself for the fitness and survival of the species. It differs from how we usually understand altruism, which means that an action is altruistic only if it is done with the conscious intention of helping another being. There is no such requirement in the biological sense.

[5] Feeling sorry for another being as a result of perceiving the distress of that being, but furthermore having the desire to alleviate this distress.

[6] A gift of self with no expectation of reciprocity.

empathy and linked to reflective consciousness. The bridge between the two levels is compassion.[7] With this division, I believe it is clear that empathy is not only a human capacity.

Through the whole of evolution, species developed the capacity to recognize and respond to the needs of others. At an early stage of evolution, recognizing distress and suffering in others was merely emotionally centered and involuntary. It promoted survival of the species and developed defensive and protective strategies. In time the capacity of emotional empathy (emotional contagion and biological altruism, led to the development of true morality in humans which clearly involves reflective consciousness. "Neural circuits established for one purpose [survival] can be exalted (exploited, recycled, redeployed) without losing their original functions" (Decety 2014: 128). In other words, high-level empathy still includes those brain function correlating with affection, but also includes developed moral cognitive brain structures. The brain structures underlying low-level empathy necessary for survival are said to be "basic neural autonomic, and endocrine systems subserving attachment-related processes, which involve the brainstem, the preoptic area of the thalamus, and paralimbic areas, as well as the autonomic nervous system" (Panksepp 1998; Decety 2014: 131). According to Jean Decety, many of the "basic components of empathy, such as emotional contagion, empathic concern and caring for others, are deeply rooted in our evolutionary history and their neural and endocrine mechanisms are highly conserved across species" (2014: 134). Hence, it seems that, even though not explicitly expressed, Decety also thinks in terms of a lower and a higher level of empathy, an affective lower and an affective-cognitive higher level.

Nevertheless, the high level is layered on the lower one. Indeed, "high-level cognitive abilities such as executive functions, language and theory of mind, which are largely implemented in the prefrontal cortex, are layered on phylogenetically older social and emotional capacities" (Decety 2014: 134).

In a previous study on intentional versus accidental actions that caused harm and damage to people and objects, with participants aged between 4 and 37, Decety concluded that "the pattern of developmental change in response to the perception of harm is indicative of a gradual shift from the affective and somatovisceral responses in young children to a more cognitive evaluative level associated with executive control of higher-order emotion and moral judgement in older participants" (2014: 139).

Perhaps, my idea of putting compassion as the bridge or link between low and high level empathy is not such a bad idea after all. However, not all neuroscientist regard empathy as having two levels. A later theory states that empathy can be explained by one neural system, the mirror neuron system.

---

[7] Feeling sorry for another being as a result of perceiving the distress of that being. Compassion is the bridge towards selfless altruism.

The Mirror Neuron System

While perhaps most philosophers think about mindreading and empathy in terms of analogical inference, social neuroscientists present a new theory for how empathy works. This theory is known as the mirror neuron theory (MNT). Most mirror neuron theorists disallow any form of theorizing, they disallow the principle of analogical inference to other minds. They argue that the discovery of the mirror neurons provides us with scientific evidence that falsifies this philosophical principle for mindreading. In my opinion and as will be shown below, it is doubtful as to whether it is possible to entirely escape from that principle.

According to the mirror neuron theorists, the fact we act empathically (regardless of the motive behind the act) is due to an underpinning *mirror neuron mechanism*. It is argued that the same neural structures are active when we detect actions, sensations and emotions in others as those that are involved in the processing of our own actions, sensations and emotions. For example, from a neurological point of view, when S sees A being sad, S is forced to act empathically, because S's neurons fire *as if it were* S who was sad.

The Neurology of the Mirror Neuron Theory

The concept of mirror neurons was coined by neuroscientist Giacomo Rizzolatti and his team in the early 1990s during visuomotor experiments with Macaca Nemestrina (1996: 131–141). The experimenters observed, to their surprise, that some neurons fired both when the monkey executed a motor act (grasping an object) and when the monkey saw another monkey or an experimenter grasping an object. According to Marco Iacoboni, due to these properties it is almost as if the monkey observed his own actions reflected by a *mirror* (2009). The neurons belonging to the neural area F5 or the Ventral Premotor Cortex (VPC) were hence named mirror neurons. Since then, the mirror neuron theory has been said to support much more than action and action recognition, e.g. action understanding, intention, etc. (di Pellegrino et al. 1992; Gallese et al. 1996; Rizzolatti and Craigtero 2004).

But according to Gregory Hickok, there is no evidence "from monkey data that directly tests this theory and evidence from humans makes a strong case against the position" (2014). Of course, even if there is no evidence for action understanding in the monkey data, it does not mean that there cannot be 'motor understanding'. Our bodies do understand bodily action in some way, but that is not the same as understanding what the action is about. One problem is that mirror theorists interpret action understanding differently. In order to be low-level empathic, it is not necessary to understand the empathic action, on the contrary, it might be better not to have to reflect on the action, since it can be a matter of survival. For example, animals may act biologically altruistically for the survival of their offspring. Empathic behavior in humans is far more complex.

Several imaging studies on humans showed that the human mirror neuron system comprises, in addition to the VPC, also the part of the Inferior Parietal Lobule (IPL) also referred to as PF or 7B; the Posterior and Inferior Frontal Gyrus, also referred to as Broca's area; and importantly, the Superior Temporal Sulcus (STS), which is not a mirror neuron system in itself.[8]

Some, but not all, maintain that the mirror neuron mechanisms are in place at birth – they are innate. However, since there is far from consensus that mirror neurons are in place at birth, perhaps it would be better to say that the preconditions for mirror neuron mechanisms are innate. Indeed, Christian Keysers and David Perrett put forward a principled explanation in simple physiological terms of how mirror properties could arise in ontogeny, i.e. during the course of the development of the individual's organism, namely through Hebbian learning. Hebbian learning is seen as a fundamental mechanism that links "sensory representations of the actions of others to motor plans" (Keyser and Perrett 2004: 503–504). Simply put, "neurons that fire together, wire together" (Iacoboni 2009: 654). Hence, according to this view, mirror neurons emerge in the brain at a certain point, i.e. when a sufficient amount of learning has taken place. This learning process is said to cause the emergence of complex behavior (Rizzolatti and Sinigaglia 2008: 222).

Let me try to reconstruct Hebbian learning.

1. Person A observes A's actions (Baby observes its hands open and close)
2. Some movements lead A to grasp an object in the particular way x (Baby grasps a toy that rotates above its cradle).
3. Thus A observes x (Baby observes Baby grasping that toy).
4. Over Hebbian time, A grasps x, y, … n. (Baby grasps different objects).
5. Now A observers B grasping y (Baby observes itself grasping a toy in a mirror or observes Mother grasping a toy).
6. Because the movement tracker STS is invariant towards x or y
7. Whenever A sees B grasping action y, A → y (whenever Baby sees a grasping action of another person, Baby will imitate that grasping action).

Mutatis mutandis, the preconditions for the emergence of mirror neuron mechanism are in place.

[8] The IPL is responsible for information processing, pain and touch sensation, spatial orientation, speech and visual perception. The Posterior and Inferior Frontal Gyrus is one of the neural correlates for cognitive functions and working memory. The Superior Temporal Sulcus is known to aid biological movement tracking and is said to play an important role in mirror neuron activity, even though it is not directly connected to the F5 or VPC area. STS neurons only fire in correlation with action-perceiving and not with action-execution. *STS neurons are not mirror neurons* but are linked to the IPL, which in turn is reciprocally connected to the VPC. It is through its double connection, (to the STS, which tracks the biological movements, on the one hand and the VPC, which plans the actions, on the other hand) that the IPL (the Inferior Parietal Lobe) plays its intermediate role in the process. In plain words, it is the IPL that processes the biological and planning information in different ways. Importantly, since the parietal lobes are essential for self-other differentiation as well as special orientation, human beings are more aware than other species of the origin of the incoming information. In addition the insula and the amygdala are essential for making mindreading possible (especially low-level mindreading or low-level empathy).

However, this line of thinking contradicts the argument put forward by some mirror neuron theorists, namely, that the mirror neuron system is the basis of imitation in humans. In other words, there cannot be any imitation without a mirror neuron system. "Although lay people are often convinced that imitation is a very primitive cognitive function, they are wrong" (Rizzolatti and Craigtero 2004: 169–192). Hence, the question of what comes first is not settled.

However, perhaps in defense of the Hebbian learning hypothesis one may reflect about the behavior of newborn infants. A baby cries (should cry) at the moment of birth. Soon, when the baby gets the attention it needs (food and the warm feeling of its mother's body), the baby stops crying. Then, every time the baby is in need of comfort or care, the baby will cry until its needs are satisfied. After a while, when the baby is put in an environment with other babies and hears another baby cry, it will automatically understand that other babies are like it, and that the other baby needs something. This suggests that the cognitive 'like-me' principle precedes the capacity of imitation of behavior and that there is other neural activity that precedes mirror neuron activity. Indeed, "newborns discriminate between their own (previously recorded) cries and those of another infant. Thus, neonates' contagious crying cannot be explained by lack of differentiation between the distress of self and other" (Decety 2014: 136).[9]

Some argue though that the capacity to imitate (mimic) has to be in place at birth. Several psychologists have observed that when newborns are put in an environment together with other babies, they try to comfort other crying babies by way of motor mimicry (imitation of behavior, i.e. in this case crying) and that some even try to capture the attention of their mother in order to lead her attention to the baby in need (Goleman 1997: 13–26). Babies from 42 min to 71 h old are perfectly capable of *reproducing or mimicking* facial movements of others. They observed that the infants perfectly reproduced tongue protrusion and other facial displays modeled before them (Meltzoff and Moore 1983: 703). For instance, Evan Thompson maintains that there is today "a large amount of evidence showing that human infants possess, at birth, interpersonal body schemas for *emotional contagion* and *facial imitation*, and that these schemas underlie the development of more sophisticated empathic abilities" (Thompson 2001: 7). This is in line with Decety's argument that high-level cognitive abilities are layered on phylogenetically older social and emotional capacities. From this I suggest low-level but not high-level empathy may be innate.

In examining the mirror neuron theory I observed that it comes in at least two versions. I will call these versions the *closed mirror neuron view* and the *open mirror neuron view*. The former view is that mirror neurons are necessary and sufficient to explain all types of intersubjective relationships (including language), provided the relationship is established between the same species and furthermore, provided the subjects have the same set of neural codes and the same psychological as well

[9] Subcortical pathways connecting the brainstem, superior colliculus, hypothalamus, pulvinar, and amygdala, all of which develop very early in fetal development, mediate this affect of sharing in (Decety 2010).

as behavioral repertoire. The latter view is more of a functional view, which means that mirror neurons *attract* different more specific neural functions in order to establish intersubjective relationships on different levels.

The Closed Mirror Neuron View

In my view, Vittorio Gallese's understanding of the mirror neuron theory could be seen as an example of the closed mirror neuron view. He understands the MNT in terms of *embodied simulation* for intersubjective understanding. The theory is known as the *Shared Manifold Hypothesis* (Gallese 2001: 47, 2005: 23–48). Providing the conditions mentioned above are met (i.e. the relationship is established between the same species and furthermore, provided the subjects have the same set of neural codes and the same psychological as well as behavioral repertoire), mirror neurons will do their job. Hence, there is a direct homogeneous link between the first- and third person experiences. "Actions, emotions and sensations experienced by others become meaningful to us, because we can *share* them with them neurologically, psychologically and behaviorally" (Gallese 2001: 47, 2005: 23–48). We *simulate* others' actions, sensations and experiences and neurologically, these actions become as if they were ours.

The discovery of mirror neurons vindicates an embodied version of simulation "'a basic functional mechanism of our brain [which is] an automatic, unconscious and pre-reflective' process" (Gordon 2005: 154), i.e. a process expressed *as you do and you feel as if it were I who did and felt.* Simply put, I will not think something along the lines of 'Oh Lisa reaches for her tea; good idea, I will do so too'. My mirror neuron system reacts "unmotivated, without [conceptual] reason[ing] or purpose" (Gordon 2005: 154). Seeing Lisa reaching for her tea, I will automatically without questioning do the same. Conceptual reasoning is not necessary (Gallese 2001, 2003: 171–180; Gallese et al. 2004:396–403; Jacob 2009: 578–588). Empathy, it is maintained, can only be established between the same species, with the same neural code and the same psychological and behavioral repertoire and it is *affective*. Empathic behavior of individuals with a deficit in the mirror neuron area would not be meaningful and remain "purely cognitive in form, pale, colorless, destitute of emotional warmth" (Rizzolatti and Sinigaglia 2008: 189).

Even though the shared manifold hypothesis consists of three levels, a phenomenological, a functional and a subpersonal level, these are seen as three aspects of the same underlying mirror neuron mechanisms, which are necessary and sufficient.

This view of mirror neuron mechanisms may provide some support for why, according to Edith Stein (1891–1942) the *exchange* of emotions between first- and third-person is immediate, i.e. why empathy is affective and happens directly. However, in my view this counts only for low-level empathy, the affective but not reflected upon type of empathy.

According to Stein, there are two categories of experiences, *primordial and non-primordial* (Stein 1989). Empathy belongs to the non-primordial one. But this means that when, for example, we are confronted with another person's distress, a primordial experience of that distress is created in ourselves, but this experience is based on a non-primordial experience of sadness (Stein 1989: 7–8). Consequently, the mindreader's experience of sadness based on an empathic encounter with another person cannot be the same as the sadness experienced by the other person.

Does not that imply that human beings depart from their own mental state as a marker for understanding the other person's mental state? And does that not further imply that the principle of inference by analogy or the very-like-me principle is adapted? Furthermore, does that not suggest that, even if the affection is immediate, there is some cognitive action involved?

What I mean is that the mindreading process of the observer may trigger additional and/or other neuroactivity than the one triggered in the target agent. For example, the perception of sadness in the other person may trigger neurons that correlate with a specific memory that previously caused the observer grief. In other words, perceiving the sadness of the other triggers a previous sadness in the observer rather than the sadness of the other as if it were the sadness of the observer. Roughly, the observers' sadness is theirs; the sadness of the target agents is their own. The observers recognize the sadness of the others because it *reminds* them of their own and not because the others' sadness feels as if it were theirs. Philosophically, when I produce behavior of type x, I feel a certain way f; therefore, when a similar body does x, the behavior was probably produced by another subject – another 'I' – that feels in a similar way f.

I believe this suggests that the principle of analogical inference for mindreading has its role as well. Hence, even though the principle of inference by analogy is not without problems, can it be neglected entirely? Perhaps it can when it comes to low-level empathy but perhaps not when it comes to high-level empathy.

The experiment performed by Leslie Brothers in 1989 is a fair example of how low-level empathy works. Briefly, Brothers' study was as follows (Brothers 1989: 10–19). Firstly, the monkeys in the experiment were taught to fear a certain tone. The monkeys soon learned to avoid this tone by pushing a certain lever whenever they heard that tone. Thereafter, the monkeys were separated and could only see each other's face through a closed circuit television. One of the monkeys was exposed to the tone. The other monkey, seeing that his fellow monkey was hurt by watching the expression on his fellow monkey's face, pushed the lever that would stop the noise. This led Brothers to conclude that empathy is a capacity that is innate in primates. As mentioned above, according to me it is the capacity of low-level empathy that is innate, which nonetheless is a precondition for high-level empathy to be developed or emerge. The monkey simply acts upon its own experience of pain and avoiding pain from the sound, and does this immediately without thinking. The monkey does not think 'my brother monkey is in pain, how can I help him'? Thinking about what to do is a human thing. Humans have the capacity to act empathically or to refrain from doing so; animals do not have this choice. Therefore, following Martha Nussbaum, I would argue that animal empathy is simple (instinctive and involuntary) and therefore it works. Human empathy is complex (affective and cognitive) and might therefore be corrupt (Nussbaum 2014).

Another example of low-level empathy in non-humans, which is automatic, immediate and unconscious, is an example of the biological altruism of the sterile workers in the social insect colonies (ants, wasps, bees and termites). These workers devote their whole lives to the queen, constructing and protecting the nest, gathering in the food, etc. According to Samir Okasha, such behavior is the most altruistic, (though biological,) because the workers, since they do not leave any offspring of their own, have zero personal fitness. They devote their whole lives to increase the reproductive possibilities of the queen (2008). Hence, it seems plausible to assert that low-level empathy is not only crucial for the survival of the offspring and species, but also that it is not restricted to the human race and that it occurs in an immediate, unconscious and pre-reflective manner.

Concerning Brothers' monkeys, one could add that this type of empathy is affective (just recall the scenario of Brothers' experiment before your eyes and you will hear the monkeys scream in despair). These examples support the closed mirror neuron view, provided it concerns low-level empathy between the same species.

The closed mirror neuron view may be sufficient to explain a monkey-see-monkey-do action measured in a laboratory; however, human beings also possess the capacity of high-level empathy that is complex, cognitive, conscious and reflective. This level of empathy cannot be exhaustively underpinned by mirror neuron mechanisms. Indeed, Gordon carefully writes, "[m]irroring systems probably play a very important role in 'mindreading' by simulation" but not the only role (Gordon 2005: 155).

The Open Mirror Neuron View

While Gallese emphasizes *representative* mirroring of the other person's actions or behavior, Meltzoff primarily speaks of *imitative* mirroring of another person's actions or behavior (Meltzoff 2005, 55–77, 2007: 126–134).

As Robert Gordon observes, this difference is crucial. Indeed, if I imitate the behavior of another person, I have perceived and understood what the other person does and I try to copy that behavior, but the behavior is mine alone and not the other's that feels like mine (Gordon 2005: 152).

In the representational version my simulated behavior is not due to my own decision-making and hence it is not mine; it is not endogenous but remains exogenous, induced from the outside by observing the behavior of another person. As already referred to above, Meltzoff, amongst others, has given evidence that infants learn through imitation. The capacity to imitate remains a significant aspect of social interaction throughout life. It is even argued that most behavior, attitudes and emotional reactions to others are acquired by imitation (Zagzebski 2004: 48). Imitation though, "[does not] require […] the concept of imitation […] and is at least partly unconscious" (Zagzebski 2004: 48). This implies that we are not always aware that we imitate, especially not during infancy as shown above.

However, imitation is a capacity that develops in time. Linda Trinkaus Zagzebski argues that it seems to be the case that people strive to perfect their own nature, to become a flourishing member of one's species. This Aristotelian assumption is formulated as "I want to be like R and not like S" because "R is [a] better [person] than S" (2004: 48). This line of thinking demands the otherness of the other more strongly than argued in the closed mirror neuron view. It also demands the integration of reflective consciousness, hence it suggest that there is more to empathy than mirror neuron activity.

I do not entirely agree with the Emmanuel Levinas (1906-1995) who defends the view of the radical otherness and maintains that any attempt to grasp the other through cognitive acts or acts of empathy reduces the other to the measure of one's own consciousness. In my opinion, actions and sensations of the other can be grasped but not entirely. There is always something of the other that remains unknown. I agree with Levinas that "if one could [really] possess, grasp, and know the other, it would not be other" (1969: 83). Hence, there remains a dimension of an experience of distress of another person that is ineffable and exterior. Without going as far as he does, encountering other persons remains a situation in which the others reveal (parts of) themselves. Such revelation of the other can inspire one to imitate that other. This suggests that imitation rather than simulation or representation seems to allow for an otherness of the other.

Discussion

The following example is meant to show that the closed mirror neuron view is not sufficient to explain empathic behavior.

Suppose you are walking on the street, minding your own business when you suddenly see a person quite roughly ill-treated. What would you do? Following the closed mirror neuron view your mirror neurons will trigger as if it was you being ill-treated. However, following the same principles, your mirror neurons will also trigger as if it were you perpetrating the crime. Now what do you do? Will you help the ill-treated person, the perpetrator, or will you do nothing at all? The fact that mirror neuron activity is involved does not settle this question.

Most human beings are empathic beings. The most probable outcome will be that you will try to help the person in need or, unfortunately, as is more and more often the case in our days, you will walk away, continuing to mind your own business or at best call for help somewhere from around the corner. However, there are cases, unfortunately, where such behavior attracts others to do the same. But let's not go there.

The essential point is that in all cases, before acting empathically or before refraining from acting empathically, you will reflect. You will do some proper thinking. Where are you at the moment? Are you alone or surrounded by other witnesses?

Suppose you are an empathic person, then in the former case you might withdraw yourself in order to call the police. In the latter case you might rush out and help the poor guy. The better your physical condition, the more prone you will be to actively intervene. Another possible scenario is that the whole scene is a set-up, e.g. a street play or part of a movie being recorded. Hence perhaps your reaction will be to smile at the camera – perhaps wondering whether you are on Candid Camera. However, the point is that the *homo empaticus* will realize this. Again, empathy, at least high-level empathy, involves more than visuomotor simulation.

This was a thought-experiment on human mindreading, but there are also examples of experiments with monkeys showing that the closed mirror neuron view has its limits.

In one experiment a monkey had to observe the action of an experimenter placing food on the table and was supposed to execute the action of bringing the food to his mouth (Iacoboni 2009). Several experiments were also performed where the experimenters showed how to use tools such as picking up pieces of food with sticks and grasping them with pliers. The experimenter observed the same activity in the mirror neuron area (F5/VPC) when the monkeys ate the food with their hands (hand and mouth actions) as when they saw the humans eating the food with a fork (tool and mouth actions). In a similar experiment monkeys were trained to use pliers in order to grasp an object. Importantly, the monkey learned to open his hand in order to grasp the object instead of closing it, i.e. the monkey had to reverse his natural grasping action. Nevertheless, the mirror neurons fired in both types of action, i.e. the primary learned action of closing the hand around the apple in order to grasp it and the secondary learned action of opening the hand around the pliers in order for the pliers to grasp the apple (Iacoboni 2009). It was argued that these experiments showed that the monkeys captured the intention behind the actions (eating). Some observations can be made.

Firstly, if it is correct that there was mirror neuron activity, mirror neuron activity can be established between different species (i.e. species not having the same neural code and the same psychological and behavioral repertoire), namely, the human experimenters and the participating monkeys. That does not mean that there cannot be some similarity between the neural architectures of different species and that in some cases this similarity may even be significant. However, and once more, in my opinion this only counts for low-level mindreading capacities.

Secondly, the monkeys were exposed to visual observation of human actions during 2 months before the experiments could take place. In other words, the monkeys had an educational introduction (Jacob 2009: 577). It is quite known that monkeys are very good at imitating.

Thirdly, once again it seems that imitating precedes mirror neuron activity (Hebbian learning). Lastly, no mirror neuron activity was observed when the monkeys saw the human experimenter mimicking eating food, i.e. picking up imaginary food. Monkeys do not mime; humans do.

While Gallese's shared embodied simulation theory dictates that a link must be established between the observed and the observer which is "constituted by the embodiment of the intended goal, shared by the agent [or target] and the observer"

(Gallese 2001: 36), Gordon argues that embodiment is about interpreting the behaviors of others "under the same scheme that makes my own behavior, along with the intentions, motor plans, and gut feelings that underlie it, intelligible to me: namely, the 'intentional' scheme of reason, purposes, and *object-directedness*" (2005: 154). Following Gordon then, mirror neurons may capture the intention behind certain behavior of another person and trigger within us a similar emotional-motoric response but it does not do so directly. Instead, the brain "may 'attempt' [...] to produce in itself a like-coded [matching] *endogenous* response" (2005: 154). This is what Gordon calls "a process of analysis by synthesis" (2005: 154).

Perhaps an even more important philosophical question is whether it is plausible to maintain that the intention behind actions and behaviors is also underpinned by mirror neurons? Can it be argued that my intention to reach for an object is a representation of your intention to reach for that object?

Some mirror neuron theorists seem to make the category mistake of confusing the intention of an agent with the belief of the observer of the intention of that agent (Jacob 2009: 582). Surely, when I see a person reaching for a cup of tea, I will most probably believe that that person's intention is to drink some tea. However, suppose the person's intention is not at all to drink tea but to put the tea-cup up-side down on the table. Would mirror neurons capture the intention of such irregular behavior? No, argues Marcel Brass, when B acts in a puzzling, unexpected manner, A immediately draws upon other mindreading resources in order to revise the hypothesis about B's intentions (Jacob 2009: 582; Brass et al. 2007: 2117–2121). Indeed, when such situations occur, the mindreader draws upon the theory of mind (ToM) and understands that the intentions of the other may be different from one's own.

Philosophically, forming an intention to perform an act oneself (without executing it), is a completely different thing than believing that another person has the intention to perform an act. Following the classical definition of mindreading, only the latter would count as third-person mindreading (Jacob 2009: 580). Since the observer has to form a belief of or predict an intention of another, he needs to possess the concept of intention himself – in other words – he has to infer the other's intended act from his own repertoire, especially when it is a tricky one. Once again it seems that the closed mirror neuron view – claiming that action-mirroring is an automatic and immediate motor-mirror process in which there is no need for theorizing or inferring, seems to fall short in explaining third-person mindreading, especially empathy. As Goldman also agrees, mindreading cannot be established by action-mirroring alone. The classical view of mindreading remains relevant.

On the basis of previous findings, Leonardo Fogassi introduced a new model of *chains of logically related mirror neurons*. Fogassi showed that the set of neurons that govern the perception of an action are not the same as the set of neurons that govern the execution of the action, suggesting thereby that the function of the mirror neuron activity found in the observer is a prediction of the target's (agent's) next act rather than a resonance or direct matching of the motor-act of the other. In the chains of logically related mirror neurons model, the activated mirror neurons *anticipate* the other person's next most likely act (Fossati et al. 2003: 579, 662–667). The chain leading to the observer's predicting an action hence includes processing contextual clues (theories). Indeed, imaging studies on third-person mindreading showed enhanced activity in several brain areas that do not contain mirror neurons.

In conclusion, the mirror neuron theory, understood in its closed version may account for low-level empathy such as emotional contagion, biological altruism and perhaps also sympathy. However, it cannot account for high-level empathy such as, for example, doing mitzvah, which is about doing something good for/to another (human) being not because such a behavior is explicitly triggered by another being, but because of the human capacity to feel empathy with all there is, with life and everything in it. This type of empathy is reflective, linked to conscious behavior of which the neural correlates exceed the mirror neuron area. In these cases, mirroring, simulation, imitation, and imagination, as well as different ways of theorizing, takes place as well.

Since the open mirror neuron view allows for the capacity of imitation, perhaps mirror neurons may contribute to the explanation of why, for example, two people, regardless of gender, who lived closely together for many years, have become each other's mirror image. Just observe old couples. Not only do they seem to know what their partner is about to say, ask or do, not only do they seem to know what their partner is thinking, they act alike and they even look alike. In psychological terms, the longer and more intensely (two) people live together, the greater their automatic mirroring and synchronicity to external input.

Compassion

As mentioned above, I regard compassion to be the bridge between low-level and high-level empathy, which in turn is the core of mindreading.[10] Hence, in my view, compassion occupies a special and important place. Compassion is feeling sorry for another being as a result of perceiving the distress of that being (any being). Where emotional contagion and biological altruism is affective, compassion is both affective and cognitive. It is not only that one feels the distress of another being; one reflects upon how one could possibly help or comfort that other being; when this happens, compassion links to sympathy which means having a deep wish to help the other being. A well-developed capacity of compassion develops (if the owner allows it) into selfless altruism. Selfless altruists help other beings for the sake of the other beings only without ever even wishing a reward or a thank you. People with a well-developed altruistic capacity do not have wishes for themselves more than to be able to serve others.

A very good example of a person who had such a developed capacity is Mother Teresa (1910–1997). She wrote,

> Let us find them [the lonely and the suffering people] and, when we find them, love them. Then when we love them we will serve them. Today God loves the world so much that He gives you, He gives me, to love the world, to be His love, His compassion. It is such a

[10] See further, Runehov Anne 2016b.

> beautiful thought for us – and a conviction – that you and I can be that love and compassion (Mother Teresa 2007: 296).

The origin of the word compassion is (lat.) com + pati, meaning to bear, to suffer.

However, as everything else that passes the eyes of philosophers, compassion is not without problems. The philosophical problem I intend to discuss is whether compassionate behavior is good or bad. Therefore, I will present and evaluate two views of compassion, the anti-compassion view and the pro-compassion view.

The Anti-compassion View

The advocates of this view of compassion mean that compassion is irrational and serves as a bad guide to action. What they mean is that the *affective* component of compassion is irrational and should not be trusted as a guide to act upon. But they also argue that compassion has a false *cognitive* structure. It is meant that people do not think as they should do in relation to the other person. They should think about the other person's dignity as well as about their own dignity as compassionate agents.

For example, if you give some coins to a person begging on the street, you do not help this person; on the contrary, you insult this person by gesturing that this person is not able to provide for her- or himself, that this person is 'below' your moral level. Everyone should take charge of themselves; *wipe their own nose.* Plato (428–348 BCE),[11] the Stoics (especially Seneca 4 BCE–AD 65),[12] Baruch Spinoza (1632–167),[13] and Immanuel Kant (1724–1804),[14] are representatives for this view (Nussbaum 2008: 336).

One can understand why these thinkers viewed compassion the way they did. To them the highest values of being human were virtue, moral purpose and dignity. Of course, people will suffer, encounter misfortunes, lose dear ones, money, status, etc. But these misfortunes were all considered as secondary. One should not pay too much attention to them. Illness and death are part of life; accept it and go on. Losing status or money is unfortunate but if you mourn about it, you merely show that you depend on worldly things, which a virtuous person does not; hence shame on you.

To these thinkers, showing compassion implied two things; firstly, you communicate a loss of dignity to the person, because the person mourns about a worldly loss and secondly, you admit a loss of dignity to yourself, because showing compassion for this person shows you are not free from worldly things either.

Kant's view on compassion is that, even though compassion is a 'beautiful' act, it has nevertheless no true moral value. Kant thinks that if one performs a moral act

---

[11] Plato. 1922.

[12] http://www.stoics.com/seneca_essays_book_1.html accessed 5 Feb. 2016.

[13] Spinoza 1983.

[14] Kant 2004.

not only because of the moral laws, but because one is inclined to do so, "it feels 'natural' to do so", then this action is beautiful (Naess 1993: 67). Kant makes a distinction between moral and beautiful. Kant's moral system is built upon the idea of an autonomous, free, rational will that only wills duty. The highest criterion for moral judgment is to act unconditionally virtuously. (Höffe 2000: 176–201). Kant's idea is based upon his principle of pure reason but in praxis. It means that one should act in a way that the maxim for one's moral act through one's will may become a universal natural law (Kant, IV: 421; Höffe 2000: 180).[15]

He distinguishes between autonomous and heteronomous will for action. Emotions and sensations are inclinations and are not free (not autonomous), something which is required to be considered morally worthy. Emotions are subject to external values; they are not universal; they are heteronomous. Of course, this has to do with his distinction of reason between empirical conditioned reason, which is sensually dependent and hence external on the one hand, and pure practical reason, which is independent on the other hand. (Höffe 2000: 169). However, he does not condemn compassionate or other beneficent actions. Compassion is not *immoral*, rather it is *amoral*.

Kant distinguishes between compassion and mercy and argues that *Barmherzigkeit* (mercy) cannot be right, because having mercy only accumulates the already existing suffering. Having *Mitleid,* (compassion) on the other hand, he sees as a duty, because without it one "would not be prompted to the caritative action in which alone our duty is fulfilled" (Davies 2001: 237). Here he clearly departs from ancient Greek and Roman moral philosophy. For instance, for the Stoics, compassion is closely linked to cruelty while *clementia* (mercy) is seen as a disposition of a wise judge towards mildness in selecting penalties. In other words, "clemency [is understood] as an inclination to dispense with *rules* governing retribution, whenever circumstances indicate that the *principles* of justice have been satisfied" (Byers 2012). For instance, Seneca begged Nero to remain a merciful emperor. Hence, clementia implies a moral reform in the end of punishment rather than simple retribution. But it is not an act of compassion (*misericordia*) which is unjust.

In conclusion, for both the ancient Greek/Roman philosophers as for Kant, it is the autonomous will that is to be praised, because it is rational, virtuous and dutiful for the sake of virtue and duty only. The heteronomous will has its values, because it has been chosen freely, but it is pathological in its core. Let us now take a look at the other view, the pro-compassion view.

The Pro-compassion View

The advocates of this view mean that bad things happen to people through no fault of their own and since this can happen to all of us, we need to be compassionate. (Nussbaum 2008: 405). To return to the example of the person begging, the message

[15] My translation from Swedish.

is to give this person not only some coins, but give the person begging some notes, because, next time you could be sitting there. The message is, *I wipe your nose and when I need you, you wipe mine.*

In this view, compassion is inspired by a combination of humanity and disaster. In contrast to the former view, compassion protects and secures human dignity while increasing your own level of dignity. In other words, by helping others, you help them to retain their dignity.

Defenders of the pro-compassion view are Jean-Jacques Rousseau (1712–1778),[16] Arthur Schopenhauer (1788–1860),[17] and Adam Smith (1723–1790).[18] These thinkers follow Aristotle's (384–322 BC) definition of compassion (*eleos*), meaning that compassion is "a painful emotion directed at another person's misfortune or suffering" (Aristotle 1999 in Nussbaum 2008: 306).

In this view, compassion has three *cognitive* components:

1. believing that the suffering is sincere and not faked;
2. believing that the person does not deserve this suffering;
3. believing that the sufferer and the compassionate possess a similar psychology.

Suffering certainly is something that normally appeals to our sense of injustice. However, (2) implies that compassion will be more easily directed to good people (having no fault in their calamity) than to those who put themselves in this situation by incorrect behavior. While compassion for the good person does not need any deliberation, compassion for the wicked person does. "Compassion requires blamelessness not only on the part of its object, but also on the part of the onlooker" (Nussbaum 2008: 313). As will be discussed later, this view is not without problems.

The third cognitive component can be described in the words of Rousseau who argues that if one is not aware of one's own vulnerability and weakness, one cannot be compassionate. The rich, he says, are hard towards the poor because they have no fear of being poor. (Nussbaum 2008: 315). Hence, in order for compassion to work, there must be a community between the sufferer and the compassionate; without a sense of a common bond, one will react with indifference or curiosity at the most. This view is shared by Aristotle and Rousseau. (Nussbaum 2008: 317).

We need to rewrite the nose-wiping sentence into, *If you are blameless, righteous and like me (i.e. belong to the social and ethnic community x), I wipe your nose.* (Perhaps understood that *I, who decide to be compassionate am blameless, righteous and belong to the social and ethnic community x*).

To Aristotle, compassion is linked to pain, memory and fear. There is a mental and/or physical pain when one apprehends another person's suffering; one may be confronted with a memory of one's own calamity when apprehending another person's suffering or one may fear that this horror would happen to oneself. (Davies 2001: 234–235). For example, today a vast amount of people volunteer to help

[16] Rousseau 2004.

[17] Schopenhauer 1992.

[18] Smith 2005.

refugees. The horrors of WWII are not forgotten. Some even remember their own calamity during the war and seeing the suffering of the refugees, relive their own pain and suffering.

For Adam Smith, compassion has not so much to do with emotions per se but with imagination of emotions, following impressions of our own senses only, not those of the other person. (Davies 2001: 236). We imagine what it would be like in this person's situation and how it would feel. Schopenhauer's idea of compassion was that it should lead to a higher state of being, to a breaking down of the ego (the will as representation). Hence, compassion should lead to selfless altruism and agape.

In conclusion, the advocates of the anti-compassion view consider compassion as irrational. The affective component of compassion is irrational but also the cognitive component has a false structure. Compassion misleads us. One thinks that one helps a person but the truth is that one insults that person by one's compassionate acts. The defenders of the pro-compassion view argue that one should help but there are restrictions. Compassion has three cognitive components, firstly, believing the suffering is sincere and not faked, secondly, believing that the person does not deserve this suffering and thirdly, believing that the sufferer and the compassionate possess a similar psychology.

Discussion

The core of the anti-compassion view is that all people are equal in value. Compassionate acts disturb this equality. This view focuses on the dignity and duty of humanity. In its extreme form, this view eliminates everything having to do with emotions. Even with Seneca's notion of mercy (*clementia*), it remains an entire cognitive (rational) moral view, liberated from affection. The idea of mercy is for a judge to be able to choose a penalty that is milder than the one appointed in law. Even if being merciful means being kind to a person in need, it does not need to include affection. The reason that Seneca allows mercy is three-fold.

Firstly, it shows the strength and dignity of the merciful, i.e. the merciful do not need to inflict pain on another person. Secondly, it shows that the merciful understands human vulnerability, i.e. humans do err. Thirdly, it is socially useful because it triggers trust rather than fear. (Nussbaum 2008: 365–366). Mercy is seen as an act of good will (which is cognitive in scope), rather than an act of compassion (which is affective in scope).

The anti-compassion view avoids the problems the pro-compassion view has such as the problem of misplaced compassion, the problem of blameless suffering, the problem of social and ethnic diversity, the problem of other species. Furthermore, it avoids undeserved self-sufficiency. Since the anti-compassion view sees compassion as being wrong or amoral, compassion is not really worth discussing further. The main problem with this view is that it does not hold, except within extremely controlled societies and within extremely controlled or idealized parts of a society or in an artificial society. One reason is that its underlying trait, empathy, in its basic form (low-level empathy), is innate in the mammal kingdom (or at least should be, nature is not perfect) and this for the sake of survival of the species. This means that in real-life situations most of us do feel with the less fortunate, the ones in need.

The problems of the pro-compassion view are misplaced compassion, blameless suffering, ethnic diversity and undeserved sufficiency. I have a personal experience of misplaced compassion.

One morning some years ago, while I was waiting for the bus I could not help feeling very compassionate towards a woman sitting on the bench in the bus-stop. She was dirty, her clothes were in rags, and her only possessions were in an old plastic bag where she kept whatever that was. Then she pulled a half-eaten sandwich out of the bag, it did not look fresh I need to say. Since I had a 100 kronor note in my pocket, I gave it to her in a very discrete manner (even though there was nobody around), telling her it was for a good meal. She looked at it and threw it in my face, calling me a rich bitch, and called me by all the names of the society in hell. I was saved from more embarrassment by the arrival of the bus. I was shaky, ashamed and wished I had never met that woman. The lesson I had learned was that compassion is not about what you feel towards another person, it is more about what they want or do not want from you. Instead of relieving a person's pain, your compassionate act may actually accumulate the pain. The sufferers may not see themselves as suffering but may see you as an intruder.

The second problem, blameless suffering is a very complex problem, however an extremely important one. We are all convinced that rape is wrong, that the offender has to be punished and the victim to be helped. Nevertheless, too many times when a woman reports having been raped, she is told that she brought this onto herself, by (for example) walking alone in a dangerous street, being dressed in a provocative manner or having drunk some alcohol. Also men may become victims of rape; as a matter of fact, an article written by Hanna Rosin shows that (at least in the USA) there is no significant difference between male and female statistics on rape (Rosin 2014). The outcome is different though, very different actually.

In a report on conviction rates for sexual offences in the UK, 2011, one can read that "60 % of court proceedings in 2011 involving sexual offences resulted in a conviction, with sexual assault on a male achieving the highest rate (91 %) and rape of a female the lowest (39,7 %)".[19] According to Lara Stemple, "feminism has fought long and hard to fight rape myths—that if a women gets raped it's somehow her fault, that she welcomed it in some way" (Stemple and Meyer 2014; Rosin 2014). That this does not seem to be the case for men is according to her that violence against men is seen as aberrant. Compassion seems to be a finite resource, which it is not.

The demand for blameless calamity, Nussbaum argues, is a human demand, not an animal one. Animals simply register suffering and comfort the one in need, with no questions asked, so to speak (Nussbaum 2014: 123–150). Animal compassion is

[19] http://www.theguardian.com/news/datablog/2013/jan/11/male-female-rape-statistics-graphic accessed, 8 December 2015.

simple (instinctive and involuntary) and therefore it works whilst human compassion is complex (affective and cognitive) and might therefore be corrupt.

The third problem concerns social and ethnic diversity. When Rousseau says that the rich do not help the poor due to the difference in their social class or difference of possibilities, he does not mean that this is always the case. Sadly though, we need to admit that too often poor people, if regarded at all, are regarded as less valuable. This is similar to people belonging to another ethnic group. Unfortunately our world today gives witness of such discrimination. The question why people help the ones closest to them might in part be explained in evolutionary terms; for example, how emotion in early prehistoric stages evolved into the emotional motivation to help others within the tribe. However, in my opinion, this reason is insufficient. Environmental, cultural, religious, and educational traits are in play as well, as are traits such as greed, pride, self-sufficiency, etc.

For instance, a study on beliefs about the causes of poverty by Matthew O. Hunt shows that belief of what causes poverty is culturally related. Indeed, in comparing the beliefs about the causes of poverty of Black, Latino and White individuals, O. Hunt observed that Latinos are more likely than whites to view both individualistic and structural explanations for poverty as important causes (Hunt 1996). Hence, if Hunt is correct, what one believes to be the cause of the poverty of one's compatriots will influence whether or not one will feel compassionate, which in turn will influence your decision whether to help or not.

With this we have come to the last problem I want to highlight, namely undeserved sufficiency. People become more and more satisfied with 'feeling' compassionate, without actually doing something that might cause them difficulties and sacrifice. (Nussbaum 2008: 399). We see this on a daily basis. Take for example all the charity programs where people are encouraged to donate some money for the cause. Also, how many times are we encouraged to 'like' and 'share' certain pages on Facebook? Furthermore, we are confronted with a variety of advertisements of charity organizations showing us how miserable our world is. Exhausting this vulnerable human feature might unfortunately lead to undeserved self-sufficiency (I have donated x money) or to neglect (I find it difficult to choose what to support).

The affective component of compassion thus seems to play an important role and hence it needs to be neglected (anti-compassion view) or restricted (pro-compassion view). For example, emotions play an important role when deciding whether people brought their calamity upon themselves or not (e.g. the 'poor' raped man and the 'careless' raped woman). We also see emotions acting when people donate to a case close to their heart (e.g. cancer research) but walk by a person begging on the street. Perhaps Daniel Batson's distinction between empathy-induced altruism and moral motivation can be of help to explain why humans will show compassion in some cases while not in others (Batson 2014). According to him, only when empathy-induced altruism (the affective component) and moral motivation (the cognitive component) cooperate, is the Golden Rule applied.

The Golden Rule is understood in two ways; a direct and a cautionary one. The direct way dictates that one should always treat others in ways that one would like to be treated. The indirect way says that one should NOT treat others in ways that one would NOT like to be treated. If empathy-induced altruism and moral motivation are in conflict, people may neglect or resist helping others (Batson 2014: 43–58).

In reality this might imply that for a male judge having to decide whether to convict a perpetrator who rapes a man, these psychological functions cooperate and the perpetrator gets his/her verdict. The male judge might feel emotionally closer to the victim, because he has the same gender as himself. In other words, he can put himself in the victim's mental shoes easily. This is due to the Theory of Mind However, if the victim is a woman, a species of the other gender, these psychological functions might (but need not) conflict and the perpetrator might go free. It is similar when compassion concerns social and ethnic diversity. You will easily feel compassion for a colleague you work with every day, because not only have you got to know this person but you also share experiences. You almost know that person as you know yourself. It is different when the calamity concerns people living in another part of the country and with whom you have never shared anything. However, suppose these people are researchers like you, you might feel and even show compassion, because, once again, you recognize yourself. To put it in the words of Nussbaum, "humans may show true compassion in cases that emotionally remind them of things that matter for them" (Nussbaum 2014: 123–150).

In conclusion, compassion is a tricky issue. The trickiest part seems to be its affective component which (under normal neurological preconditions) cannot be avoided. The anti-compassion as well as the pro-compassion view tries to avoid or limit the influence of emotions. The anti-compassion view does this by ignoring compassion all together, while the pro-compassion view accepts compassion but in a restricted manner since they claim worthy compassion. In reality, this does not work. Inspired by Batson, I would say that when these components cooperate, the Golden Rule may be applied; however I do not say that this will always be the case. Besides neurological preconditions (having the ToM and no neural deficiencies), factors such as culture, religion, education and politics seem to want to have a voice as well. Furthermore there will be the factor of motivation. If a person after watching a charity show on television picks up the phone and donates some money to that charity, what is happening? Is this an act of compassion? An act of obligation (this is something we have learned we all have to do)? Is it an act of bad conscience (not having lived up to the expected moral norm)? If a person decides to leave a comfortable home and country to contribute practically in a charity movement in some third-world country, is this an act of compassion? Is it an act of fulfilling a wish for adventure? Is it an act of escaping dullness and boredom? If a priest visits the sick, is this an act of compassion? Is it an act of duty? Is it an act of faith? Many scholars and scientists have studied compassion. Nevertheless, I believe we are only seeing the tip of the iceberg. The question is what lies underneath?

Conclusion

This chapter was concerned with the capacity of mindreading with emphasis on the capacity of empathy and compassion. Mindreading is simply the capacity humans have to understand other minds. Empathy, it was argued is the core of mindreading. However, empathy as is compassion is a tricky concept. In order to simplify the concept of empathy it was divided into low-level and high-level empathy. Low level empathy was argued to comprise emotional contagion and biological altruism which is automatic and imitative and voluntary and merely affective High-level empathy was argued to comprise sympathy, altruism and agape which are more complex, also affective but highly cognitive and are hence linked to conscious reflection. As a bridge I established compassion.

That there seem to be two levels or orders of empathy, is supported by neurological studies a more automatic, imitative level and a more complex cognitive level (Seitz et al. 2008) or an emotional level of empathy necessary for survival and a more cognitive level as developed in humans (Decety 2014). However, since the discovery of the mirror neurons it has been argued that this neural area is enough to explain empathy.

I have tried to show that such a conclusion is not tenable by way of looking at two different understandings of the mirror neuron theory, which I called the closed and the open mirror neuron view. My suggestion was that the mirror neurons may be responsible for the capacity of low-level empathy but when it comes to high-level empathy more neural functions are involved, not least functions in the frontal lobes, which support the cognitive level of empathy. Therefore I also suggested that the philosophical idea of analogical inference cannot entirely be excluded, certainly not when it comes to high-level empathy.

Compassion was defined as feeling *sorry for another being* as a result of perceiving the distress of that being. Compassion is hence different from low-level empathy in the sense that it does not merely concern perceiving (feeling) a distress of another being (and in this case a kin), it concerns feeling *sorry* for that being (not only a kin). Feeling the distress of another being is an emotional experience; feeling sorry for another being is both an emotional and a cognitive experience, in the sense of understanding what the distress is about. However, compassion is not always seen as a psychological trait worth endorsing.

Some defend the anti-compassion view while others promote the pro-compassion view. The advocates of the anti-compassion view argue that compassion is irrational and serves as a bad guide to action; hence, get rid of your compassionate emotions. Their view is based on the idea that life is filled with calamities of different kinds; keep your head up and go on. Dignity is what matters. The message is *you wipe your own nose and I'll wipe mine.*

The advocates of the pro-compassion view also base their view on the idea that life is filled with calamities; however, they also emphasize that calamities may be disastrous for the person undergoing calamity. Furthermore, they realize that such calamity might happen to them as well and as a result make their lives disastrous.

Hence the message becomes *I wipe your nose and when I need you, you wipe mine.* Nevertheless, there are restrictions on when it is permissible to wipe another person's nose. Be sure that your compassionate act is not misplaced, be sure that the suffering person did not bring the calamity onto himself through his own fault and do not waste your compassion on those who are not part of your social or ethnic group. The message becomes, *if you are blameless, righteous and like me, I wipe your nose.* (Perhaps understood that *I, who decide to be compassionate am blameless, righteous and belong to x social and ethnic community*). It is needless to say that such limitations are problematic from a democratic moral point of view.

It seems that both the topic of empathy and perhaps especially compassion need more research.

It is now time to investigate situations where empathy and compassion do not prevail. Hence the philosophical problem of free will and evil is revisited.

Chapter 6
Free Will, Responsibility and Moral Evil

Introduction

While the previous chapter considered empathy and compassion, this chapter looks at some reasons for why empathy and compassion sometimes fail to prevail. What could be the reasons that some dismiss or even lack empathy? Furthermore, where does free will come in? Is there something like free will or is free will always relative to a specific factor? Is free will equal to being free? In my opinion having free will and being free is not entirely the same. For instance, a drug addict may have freely chosen to start taking drugs, but after some time it is the drug that chooses the addict who is no longer a free person. Take the following example. I am completely free to choose what I want to do today, I may choose to continue writing on my book or I may choose to do something entirely different. However, I am not free from my duty to submit the book in time. Then, of course, one could try to refrain from making a choice, thinking in terms of 'what has to happen also happens'. However, that will not work, one will end up making a choice because "[w]e cannot think away our free will" (Searle 2007: 43). However, that we cannot think away our free will does not mean that we have absolute free will. In turn, this does not mean that free will is an illusion. As Searle expresses, "free will is a genuine feature of the world" (Searle 2007: 58). Therefore, it needs to have neurological correlations. The choices we make cannot be detached from our neurological set-up. This implies that human free will to do good or bad is at least partly hardwired in the brain.

To repeat the question, do human beings possess genuine freedom of will? If that is the case, is such genuine freedom of will compatible with predetermined cultural, psychological and neurological properties? If there is no such thing as freedom of will, how then could people be held responsible for their moral actions? This chapter accounts for three philosophical views, namely, libertarianism, incompatibilism (causal determinism) and compatibilism (voluntariness and self-determination).

The conclusion is that indeed, human moral acts are determined by their culture, psychological and/or neurological condition. However this does not mean that

A.L.C. Runehov, *The Human Being, the World and God*,
DOI 10.1007/978-3-319-44392-8_6

people cannot be held responsible for their moral actions. Rather, it means that there seems to be a scale of responsibility corresponding to which types of moral evil are performed. Hence, four types of moral evil are suggested. The first type of moral evil I have called *pure accidental moral evil.* The second type I called *belief-based accidental moral evil.* The third type I have named *active but not entirely-responsible moral evil* and for the last type I have chosen the term *active and responsible moral evil.*

The Problem of Evil

In order to draw attention to the importance of understanding the complexity of moral evil acts, different moral evil acts with which most of us are acquainted are presented. The first one refers to the 7/11 terror attacks on the twin towers in New York. What happened that day in 2001 had, as one of the many consequences, the effect that the opinion on religion for many people changed. For instance, Dawkins' response to the terror attacks was,

> Many of us saw religion as harmless nonsense. Beliefs might lack all supporting evidence but we thought, if people needed a crutch for consolation, where's the harm? September 11th changed all that. Revealed faith is not harmless nonsense; it can be lethally dangerous nonsense. Dangerous because it gives people unshakeable confidence in their own righteousness. Dangerous because it gives them false courage to kill themselves, which automatically removes normal barriers to killing others. [...] And dangerous because we have all bought into a weird respect, which uniquely protects religion from normal criticism. Let's now stop being so damned respectful (Dawkins and Richar 2001).

Indeed, terrorist actions have been, and are, one of modern civilization's biggest problems. Furthermore and unfortunately, often it is claimed that the acts were performed in the name of God, in the name of divine justice.

Friday the 13th (a date people associate with evil and bad luck) lived up to its reputation on November 2015, when a series of coordinated attacks occurred in Paris. 130 people were killed including 89 at the Bataclan theatre; 368 people were wounded, 99 seriously. Seven terrorists lost their lives as well. Similar to the terror attacks in New York, the perpetrators belonged to a religious group. January the same year, the French satirical weekly magazine *Charlie Hebdo* was attacked by a similar religious group, killing 12 journalists. We also remember 11th March 2004 Madrid train bombings where 191 people were killed and more than 1800 injured. Once and unfortunately again, the terror attack was performed by religious extremists. A last example of such terror attacks is the 7th July 2005 London bombings (7/7). Similar to the terror attacks in Paris last year, the horrible event consisted of a series of coordinated attacks on central London's public transport systems during rush hour. 52 were killed and more than 700 injured. The question is, would the world be better off without religion? Would we live in peace and harmony if there were no religions at all as John Lennon sang?

Consider the following example. On the 7th November 2007, eight people, seven schoolchildren and their headmaster, were killed during school-time in Tuusula in Finland. The perpetrator was an 18 year old student who unexpectedly started shooting his gun during lecture hours, having failed to set the school on fire. He had made his manifesto available on the internet some weeks before his action. He wrote,

> Life is but a meaningless moment, a long evolution and result of many different facts and causes. However, life is also, what one, as an individual, makes of it. And I am the dictator and God of my own life. [...] I have prepared myself for this camp and prepared myself to die for my thoughts. As a natural selection, I will kill all those I think are improper, all those who are a disgrace for humankind and all those who should not be permitted to live. [...] You will probably say that I am insane, a psychopath or criminal, but the truth is, I am but an animal, a human being, an individual [...] I am ready to die for my cause, the cause I know is the right one, just and true (Aargaard 2007).[1]

The perpetrator's source of inspiration seems to have been the idea of natural selection and the survival of the fittest or in other words, the destruction of what he considered the weakest, unfit for evolution.

I believe everyone will remember the Norwegian Anders Behring Breivik, a far-right-wing supporter who killed eight people by setting off a bomb before the government quarter in Oslo, and then murdered 69 participants of a Labor Youth League (AUF) summer camp on the island of Utoeya on the 22nd July 2011. Yet another example of non-religious evil I want to refer to occurred in Sweden, in a little village called Trollhättan. Also in this case the motive of the perpetrator was racial. He entered a school carrying a sword and a Star-Wars mask and attacked foreign born students. In his actions he killed a teacher and two boys, only 17 and 15 years old. This happened on the 22nd October 2015.

On the 12th November 2015, two suicide bombers detonated explosives in Bourj el-Barajneh, Beirut, Lebanon, causing 43 deaths and about 240 wounded. The Islamic State of Iraq and the Levant (ISIL) claimed to be responsible for these deeds. The same year, on Friday the 13th, a date associated with evil and bad luck, it became Paris' turn to be terrorized. Several explosives were detonated in Saint-Denis, a northern district of Paris. The targets, i.e., the Stadium, cafés, restaurant and a music hall, were crowded with people. 130 people died, 368 were injured, about 99 severely. Also this time, the Islamic State of Iraq and the Levant (ISIL) claimed to be responseble for these deeds. They also claimed to be responsible for the terror attacks at the airport of Brussels, and the underground nearby the EU buildings, on the 22nd March 2016. 35 people died, 316 were injured, 62 seriously. Only some days later, on the 26th March, suicide bombers entered a football stadium during a football match. The perpetrators, claiming *responsibility* belonged to the IS, and killed 41 young people and injured 105. A last and recent terrorist attack occurred on the 28th March, in Lahore, Pakistan, where suicide bombers entered Gulshan-e-Iqbal Park, one of the largest parks in Lahore, targeting Christian families who celebrated Easter. The majority of the victims were women and children,

[1] My translation from Danish.

both Christians and Muslims. 75 people died, 340 were injured. Jamaat-ul-Ahrar, a group affiliated to the Pakistani Taliban, claimed to be responsible for the attack.

Perhaps we should reiterate what is asked above, would the world be better off without worldviews, regardless of whether these are religious or secular? Obviously, the way in which people comprehend the world, i.e. how people have constructed their worldview, plays a most significant role in the way in which people morally treat others (i.e. their human fellows, animals, as well as the world's flora). While some blame the moral evil actions in the world on worldviews, others blame these actions on evolution.

Similarly to worldviews, evolution cannot be excluded from the overall understanding of moral evil. However, to say for example, that the origin of contemporary people's immoral acts is to be found in the predation of our hominid ancestor is problematic. One reason is that we lack sufficient scientific evidence in order to reliably support such a view. (Zangwill 2006). From the perspective of evolution, moral evil actions are related to empathetic ones. That may sound confusing; however the fact is that the suffering of one group (evil) has been shown to be necessary for the protection and survival of another (empathy). It is argued that empathy had to evolve in order to establish cooperation within the group (Proctor and James 2005). Obviously, emotionally understanding one's fellow members implies that one will take action to help one's fellow members when they are in need. Empathy, it is argued aims at maximizing the survival of one's own group and securing reproduction. This necessity is said to explain the evil in the world (Fieldman 2003:122–132).

A counter argument is that there is no logical relationship between our ancestors killing for food or for other necessities for survival and the world's contemporary terror. Hence, neither worldviews (cultural-religious-personal understandings of the world) nor the evolution of the human species can be entirely excluded from the explanation of why there is moral evil in the world. I believe that the problem of moral evil is far more complicated than that.

The complexity of the problem of evil is related to the human capacity of being *moral agents*. The problem of moral evil is also related to the problem of free will and hence also related to the problem of responsibility. The questions longing for an answer are, whether human beings possess genuine freedom of will as some philosophers argue. If that is the case, is such genuine freedom of will compatible with predetermined cultural, psychological and neurological properties? If there is no such thing as freedom of will, how then could people be held responsible for their moral actions?

Moral Evil

Sometimes evil is defined in terms of a noun, in other words as *something, a fact, an evil force* that causes suffering and misfortune. An example of noun-based evil would be evil embedded in the order of nature such as natural death and natural catastrophes, which every second hits humans, animals and nature alike. Such evil is ranged under the term of natural evil. Natural evil, it is argued has no intention; it just happens and happens to all.

However, evil is also defined as an adjective, as a cause of being morally reprehensible. This kind of definition deals with people's moral wickedness. Being morally reprehensible then is said to be caused by bad character traits, on the one hand, and bad qualities of a natural kind, on the other hand. Evil concerns anything which impairs the happiness of a being or deprives a being of any good. The *anything* refers to both natural and moral evil. Wickedness, crimes, violations of laws and rights are seen as moral evils. Natural catastrophes, on the other hand, are classified as natural evil. One consequence is that people are considered to be responsible for moral evil, but not for natural evil. In my opinion, the line cannot be drawn that sharply.

Are not natural catastrophes, at least in part, also a consequence of how human beings treat nature? For example, the floods in parts of Europe and America are at least partly due to the expansion of infrastructure, agriculture and buildings. In addition to this, the global warming problem is at least partly caused by modern human lifestyle. This means that natural evil caused by human actions or the lack of human actions also belongs to the category of moral evil, even though human beings may not have intended to cause evil. After all, what humans developed during the agricultural, the industrial and modern era was not intended to harm; rather the aim was to enhance life. Nevertheless, it seems that the problem of evil is far more complicated than some will have it. Dividing this problem into moral evil issues and natural evil issues is not enough.

One who has contributed with fruitful further distinctions is Thomas Anderberg (1956–2013). He divides moral evil into accidental moral evil, which is non-intended moral evil and active moral evil, which is intended (Anderberg 1997: 35).

Moral evil refers to (1) the way in which moral agents act (adjective) as well as to (2) the moral agent's acts in themselves (noun). Furthermore, we deal with both intentional or active and accidental moral evil. One consequence of this reasoning is that it seems to be that moral evil, understood as adjective, also includes activity (or intentionality) while moral evil understood as the act itself does not necessarily do so. Moral evil understood as an adjective expresses a motion while moral evil understood as a noun expresses a result. Perhaps why this is so may be explained by how freedom of will is commonly understood.

Commonly, people have the idea that human beings are the only creatures possessing freedom of will. In other words, common sense tells us that human beings are free to perform or refrain from a certain action at any time. In other words, it is argued that human beings do things or refrain from doing things because they want or do not want to do these things. Consequently, it is claimed, human beings are always *responsible* for their actions, because they *intentionally* perform or refrain from acting.

One philosophical observation is that the concept of free will (the freedom of will) actually consists of two value laden words, freedom and will. Furthermore, there is no consensus on the relationship between the concept of freedom and the concept of will. Some philosophers maintain that possessing freedom means possessing the capacity to make decisions and to act or refrain from acting accordingly (this is similar to the commonsense view). Other philosophers argue that possessing freedom to make decisions and possessing freedom to act are two different things. Possessing freedom to make decisions is, according to them related to the human will (to do or refrain from something) while possessing freedom to act does not need to have a relationship with wanting (willing) to act or not. In other words, in contrast to acting, making a decision typically involves intentionality (Pink 2004: 2–7).

While there are problems with this view, I believe that making at least a semi-distinction is fruitful because, as will be clarified later on, there is a difference between a person who *wills* something (and acts accordingly or not) and a person who does something *without* actually *willing* it.

As mentioned above, a person may accidentally cause harm. Furthermore, people may have good intentions (building highways in order to facilitate transport) but the results or the acts in themselves may be bad (natural catastrophes such as floods and earthquakes). Also, a physician intentionally causing pain to a patient may lead to a diagnosis, which finally cures the patient. I believe we begin to have a clue of the complexity of the problem of evil. How people act has moral implications, regardless of whether the acts were intentional or not.

The question is when do people become responsible for their acts and to what extent are people responsible for their acts? Are human beings always responsible for their deeds as, for example, Immanuel Kant, amongst other philosophers defending libertarianism, believed them to be? Let us take a look at two different views of how freedom and responsibility are related, libertarianism on the one hand and incompatibilism on the other hand.

Libertarianists believe that human beings possess total control over how they act, i.e. human beings are free agents. It is this type of freedom that is considered to be a necessary condition for moral actions. On the other hand, advocates of incompatibilism or causal determinism argue that human moral actions depend to a large extent on the environment in which they are born, their genes with which they are born, their desires and feelings that come over them beyond their control, etc. (Pink 2004: 13). These philosophers claim that freedom is incompatible with the causal predeterminations. One question I want to raise is which view would be most adequate in the age of neuroscience, libertarianism or incompatibilism?

According to Kant, moral evil as such does not exist. He argues that it is "[m]an himself [who] must make or has made himself into whatever, in a moral sense, whether good or evil, he is or is to become. Either condition must be an effect of his Willkür [will]; for otherwise he could not be held responsible for it and could therefore be morally neither good nor evil" (Kant 1960: cxi). Hence to Kant, the moral quality of the will is completely self-acquired. In other words, humans will good or evil. This notion of evil Kant calls *radical evil.*

Kant's radical evil is understood as the counterpart to the Christian doctrine of original sin. Radical evil, in contrast to original sin (which is understood to be inherited) is self-incurred by each human being and hence, every human being is responsible for this radical evil and moral conversion (2014).[2] Furthermore, before a human being makes a decision to act rightly or to act wrongly, the human being is innocent. Kant writes, "[f]or whatever causes were to be found within him or outside him, his action is yet free and determined by none of these causes" (Kant 1960: cxii). How then does a human being become corrupt? According to Kant, the source of moral evil is to be found in human beings' "tendency or disposition to will the rejection of himself as a self-determining personality, as a free being, for the sake of himself as a creature of nature" (Kant 1960: cxii). Hence, if Kant is right, there is no good or evil (morality) in itself; it is the human being, disposed to act in one way or another, that is responsible for the consequences of his or her acts, regardless of which conditions might underlie these acts. For Kant, something is good or evil if the will, which created it, was good or evil.[3] In other words, in a Kantian sense, to call a thing evil, in a moral sense, is to call it an act of somebody's will. According to Kant, God gave human beings free will but many choose to abuse God's gift. Consequently, in order to overcome radical evil, human beings have to reorder their fundamental principle of choice.[4]

Kant's definition of moral evil could be compared to Anderberg's definition of active moral evil. The Kantian view suggests that the problem of goodness is the problem of the good will; hence the problem of evil becomes the problem of the evil will. This is so because Kant emphasizes that free will is given to all human beings (by God). Freedom and will belong together. I found this view philosophically problematic, because such an understanding of moral evil does not reflect the possibility of accidental moral evil as mentioned above. The libertarian view (which similar to Kant emphasizes the existence of human free will) does not pay enough attention to cultural (for example, educational), and evolutionary (for example, neurological) aspects.[5] We could ask whether, the Finnish, Norwegian and Swedish perpetrators mentioned above, as well as the members of religious and ideological groups causing terror are victims of their culture, education, and/or neurological make-up, as would be claimed by the advocates of incompatibilism (or causal determinism)?

Another problem with Kant's view is, if something that ought to be considered as good or evil is due to the will (intention) creating it, how should we understand that good intentions sometimes cause calamities and, on the other hand, that bad intentions sometimes turn out to have favorable consequences? For example, giving money to a beggar were instead of buying food or shelter with the money, the beggar buys liquor and finally dies. An example of a bad will resulting in something good

---

[2] http://plato.stanford.edu/entries/kant-religion/accessed 5 February 2016.

[3] The difference between Kant's moral philosophy and Utilitarianism is that Kant sees the good will as an end while Utilitarianists see the good will as a means, i.e. in a derivative sense.

[4] Kant does not believe in the doctrine that Jesus Christ took away the guilt of previous evil (vicarious atonement) but argues that we should see Christ's actions as an example (exemplary atonement).

[5] Perhaps this note is unnecessary but Kant did not have access to psychological or neuroscientific studies on moral behavior as we have today.

would be a boss's ganging upon an employee and finally sacking the employee without any firm reason resulting in this employee finding a better job.

However, the view that there is no such thing as free will (causal determinism) is also problematic because, if there is no such thing as free will to act or refrain from acting, would that not imply that human acts are merely caused by chance? Furthermore, as Pink asks, "would human acts be acts at all or merely blind motions" (Pink 2004: 16).

While libertarians need to explain in what sense causally undetermined free acts really are free, compatibilists (causal determinists) need to explain in what way human beings could be held responsible for their acts presupposing everything is predetermined. Obviously there are circumstances in which people cannot be said to be entirely free. I want to emphasize two such circumstances. The first one is concerned with the consequences of coercive psychological methods for moral actions. The second one concerns the neural; the consequences of certain neurological conditions for moral actions.

Coercive Psychological Methods

A person may become a victim of psychological methods aimed at controlling the person's mind. Such coercive psychological methods may include brainwashing, thought-reform and coercive persuasion.

By way of suggestibility-increasing methods; sleep and nutrition restriction; humiliation; social isolation and creating economic and other dependence, individuals become unable to make rational balanced decisions and they are weakened to such a degree that they will follow whatever is ordered. Mind-control is generally known as a powerful method which, when used in medical circumstances may cure people from psychological disturbances, but when used in political, religious or other ideological settings, this powerful method may also become dangerous (Beier 2006: 301–301).

Nevertheless, libertarians could still argue that people, being free creatures, have a choice whether they want to join such groups or sects. They could argue that people actually have the possibility to investigate the situation beforehand (Beier 2006). However, that is not always the case. That such methods have been used by, for example, the German Democratic Republic is well documented. Leaning on the US Department of State's *"Country Reports on Human Rights Practices 2004"*, Hermán Reyes calls such coercive psychological methods psychological torture (2007). Various psychological methods are described as torture,

> [M]ethods of torture include … prolonged periods of exposure; humiliations such as public nakedness; confinement to small "punishment cells," in which prisoners were unable to stand upright or lie down, where they could be held for several weeks; being forced to kneel or sit immobilized for long periods (Reyes 2007: 598).

Coercive psychological methods have been (and are still) used to break down people's willpower, to make them confess (whether or not they are guilty of the accusation) or to make sure they will remain loyal to the *party*. It is needless to say that the use of psychological torture has been shown to have destructive health consequences on the people undergoing it. The testimony of victims is often imprecise concerning time, location and details, and often the victims unconsciously or deliberately forget about it all. (Reyes 2007: 601). Hence, the question whether victims of coercive psychological methods can be taken to be responsible for their moral acts is far from decided upon.

Another type of circumstances that makes it difficult to agree entirely with the libertarians is neurological in scope.

Neurological Underpinnings

Let us briefly return to what has been said above concerning empathy.

Daniel Goleman maintains that we usually recognize anxiety in another person's voice unconsciously (Goleman 1997: 97–98). Empathy, he argues, is a human emotional capacity already visible at birth. Empathy is the capacity to understand how other people feel. Even if there might not be a real consensus on how to define empathy exactly, the following elements include; (a) The capacity to affectively respond to another person. (b) The cognitive capacity which means that one can put oneself in the other person's condition. (c) The capacity to regulate one's emotions in relation to the other person (Decety and Jackson 2007). Remember the example of how babies try to comfort other crying babies by way of motor mimicry (imitation of behavior) as well as by leading their mother's attention to the baby in need. Also, that at two years of age, infants have the required cognitive, affective and behavioral capacities to display concern for others (Bretherton et al. 1986: 529–548). Furthermore, there is more than enough evidence that infants already have interpersonal schemas for emotional contagion at birth and that these schemas underlie the development of more sophisticated empathic abilities (Thompson 2001: 7). Importantly, psychologists also argue that children whose parents pay attention to their child's (mis-)behavior, have a greater capacity for empathy. Indeed, attunement is the determining factor in how much empathy a person will develop; children need to feel well attuned to their parents or parent substitutes. That is not to say that early misattunement cannot be corrected through, e.g. other relationships later in life.

However, when children feel that they are not loved, not cared for and feel abandoned, they *may* take revenge for this neglect on their parents, parent substitutes and society. These children will have a lack of empathy, or rather, a not sufficiently developed capacity of empathy. It has been suggested that this may have been the reason behind the massacre in Finland mentioned above.[6] Note that a lack (or loss) of empathy may lead a person to behave immorally but there is no direct logical relationship between the two parameters.

---

[6] Of course, a lack or loss of empathy and an increase of violence may also be caused by drug abuse.

If it is the case that the brain is designed in such a way that the preconditions for empathy are indeed hardwired in the brain, then we could expect a person whose empathic neural preconditions are not well developed or not developed at all, to have a lack or loss of empathy. Several neurological studies show that such is the case.

It has been shown that lesions of the *amygdala* occurring in early development and which act as *epileptogenic foci* in early childhood caused deficits in *Theory of Mind (TOM)* reasoning (Shaw et al. 2004: 1535–1548).[7]

Other studies showed that damage to the amygdala makes people unable to recognize fear from facial expressions as well as the phenomenological experience of fear and/or from body postures and emotional sounds (Adolphs et al. 1995: 5879–5891; Sprengelmeyer et al. 1999: 2451–2456; Runehov and Anne 2010). These persons neither recognize anxiety on another person's face (which Brothers' monkeys did recognize) and neither do they understand that another person might be in pain when that person screams (which, normally according to Goleman, newborns understand). Neuroimaging studies on the perception of pain in others also show that there is a difference in amygdala activity when evaluating the pain of others compared to evaluating one's own pain (Decety and Jackson 2007: 93).[8] In other words, one is aware of one's own pain but unaware of the pain suffered by others. Studies showed that "offenders with a high level of psychopathy show reduced autonomic arousal while viewing a confederate receiving electric shocks" (Aniskiewicz 1979 in Decety 2014: 142). When the amygdala is stimulated electrically, animals respond with aggression. When it is removed, animals get very tame and no longer respond to things that would have caused rage before (Decety and Jackson 2007: 93; Shamay-Tsoory et al. 2005: 468–472).[9]

Psychopaths are indeed neurologically differently shaped. Brain abnormalities that are mapped are, "reduced volume of the amygdala, reduced gray matter volume in the frontal and temporal cortex, especially in the right posterior superior temporal sulcus, and in the insula, and increased volume of the striatum" (Decety 2014: 143).[10] This means that the brainfunctions correlating with empathy are reduced in psychopaths, while the brainfunction correlating with reward is increased. However,

---

[7] The amygdala which is located in the middle part of the temporal lobes is one of the oldest structures in the brain and controls and mediates virtually all high-order emotional functions. However, as argued above, one should be aware of the fact that no neural activity stands alone. Amongst other neural functions the amygdala is connected to frontal lobe activity which is maintained to be responsible for higher cognitive functions. Epileptogenic focus is the area of the cerebral cortex (the outer portion of the brain) responsible for causing epileptic seizures.

[8] Also in the superior temporal sulcus, the anterior insular, and the areas of the premotor cortex. Both the cognitive (frontal) and affective (limbic) processes are interwoven in neural networks that participate in mediating the empathic response (high-level empathy).

[9] fMRI studies on the perception of pain in others also showed that there is a difference in neural activity when evaluating the pain of others compared to evaluating one's own pain.

[10] I have already shown the importance of the amygdala, the frontal and temporal cortex and the insula for empathy to emerge. The striatum, or neostriatum or striat nucleus is a subcortical part of the forebrain and very significant for the reward system.

studies have also shown that while persons who score high on psychopathy do understand moral norms, they disregard them (Aharoni et al. 2012; Decety 2014: 143–144).

While every brain is unique, (there is no such thing as two equal brains, similar to there not being two equal faces), we can conclude that some brain dysfunctions are significant for a person's level of socio-emotional responses to others; these are "the amygdala, insula and vmPFC" (Decety 2014: 144). If these neurofunctions are not functioning as they are expected to do, there will be lack (or even entire loss) of empathic/compassionate arousal. Nevertheless, this is not in itself a sufficient explanation for moral evil.

The interesting question is now, what is the relevance of neurological studies for understanding moral evil?

Neurology and Moral Action

In my view, understanding the neurological underpinnings of moral evil is crucial, because, it seems that there are people who (1) act *amorally* (they lack or have a severe loss of the capacity of empathy) or (2) act *immorally* (they act accidently or intentionally bad). On the one hand it could be argued that a least the preconditions for empathy seem to be innate (i.e. in place at birth); those who possess a neurologically normally developed and functioning brain that is not mind-controlled have the choice to be morally good or bad. On the other hand, psychological and neurological studies on empathy suggest that being good or not cannot entirely depend on an individual's free will, something that supports the existence of accidental or non-intentional moral evil but also supports the effects of psychological torture and neural dysfunctions. This casts a shadow over the view that human beings are really free.

The question is now, to what extent are human beings free and to what extent are they determent? Some argue that freedom is connected to having control. For instance, Thomas Hobbes (1588–1679) thought that freedom is but an illusion; he expressed this as follows, "[t]hat which I say necessitates and determined every action is the sum of all those things which, being now existent, conduce and concur to the production of the action hereafter, whereof if any one thing were wanting, the effect could not be produced" (Hobbes 1654, § 11). To Hobbes this was the decree of God. Hence, if we accept Hobbes's view, people's moral actions are determined by their previous cultural, religious, educational, environmental, personal and neurological properties.

However, Hobbes also makes a distinction between being free from necessity (which we cannot be) and being free from compulsion (which most of us can be). Without digging deeper into Hobbes medieval thinking, his idea is relevant. This means, at least as I see it, that whether human beings possess free will (from necessity) or not, is not really essential for moral action. What is essential is to investigate the extent to which human beings can act voluntarily and hence to investigate to what extent people can be taken to be responsible for their actions.

For example, suppose person *A* decides to take the bus instead of the car. This is clearly a voluntarily action. However, what *A* does not know is that the car is broken. This means that *A* can voluntarily take the bus but not voluntarily take the car. Hence, one could argue that not being able to voluntarily take the car is predetermined (by the condition of the car). On the other hand, the voluntary act to take the bus is still voluntarily, because *A* could have chosen to take the underground, to cycle or to simply walk to the office.

This view room for both accidental and intentional or active moral evil. With this comes responsibility. Thus, to what extent are humans responsible for their acts?

Human Actions and Responsibility

Let us return to Hobbe's idea. Could it really be the case that peoples' action are by necessity generated by their culture, religion, education, environment and neurological make-up? Perhaps this is true. The question is whether people can be held responsible for bad actions due to cultural and other circumstances. People might not be held responsible for their bad actions due to a dysfunctional neural make-up, but I dare to argue that they may be held responsible for those bad actions that result from culture, education and upbringing.

I do not mean, of course, that people can be made responsible for the environment they are part of, but they can be made responsible for how they respond to this environment. One can choose to accept or not the wickedness of one's environment. This is of course not always easy; on the contrary; but to free people entirely from responsibility for their misbehavior because they belong to a certain environment is wrong. It is different for people who become victims of harsh coercive psychological methods. The responsibility for bad actions resulting from such methods lies with those who execute the methods. However, I would dare to say, once fully recovered, these victims are again responsible for their actions. It cannot be right to continue to blame one's deeds on previous mistreatments. Again the interesting question becomes to what extent people can be made responsible for their actions.

It seems that there are at least four types of moral evil; *pure accidental moral evil, belief-based moral evil, active but not entirely-responsible moral evil* and *active and responsible moral evil.*

By *pure accidental moral evil*, I mean moral evil, for example, caused by human beings who manipulate nature in a way that this manipulation results in some natural catastrophe. Perhaps their acts should not be considered as voluntary, as long as the actors are not aware that their life style has such negative effects on nature. The same counts for those who perform bad acts due to neurological failures. A child diagnosed with, for example, ADHD, who causes pain to another child, would probably not be held responsible for its acts, nor would the acts be regarded as voluntarily performed until, the child receives medical care for its dysfunction and has learned which behaviors are not acceptable.

Belief-based accidental moral evil could, for example, be caused by victims of mind-control. Their acts are voluntary but are also determined by some kind of belief, even if that belief was forced upon them. For example, in 1997, the Heaven's Gate, an American UFO-religion was responsible for the suicide of 38 people, including children. Since the planet Earth was about to recycle they were ordered to leave it immediately in order to survive. Obviously, the members of this cult were mind-controlled. Similarly, The Order of the Solar Temple, a French right-wing political sect, responsible for 179 deaths by mass suicides and murder between 1994 and 1997; and the People's Temple founded by James Warren Jones, driving 912 people, of whom 276 were children, to commit group suicide in 1978. The people received a "deadly concoction of purple Kool-Aid mixed with cyanide, sedatives and tranquilizers".[11]

Other kinds of belief-based moral evil acts may be caused by Islamic suicide bombers, Japanese suicide bombers during WWII, people indoctrinated by, for example Stalin's or Mao's communism (Wiser 1983: 368) and Hitler's Nazism, etc. How many people blindly believed in their leader, Hitler? Those who read Mein Kampf knew that Hitler explained his motive as the will to power, in eliminating the weak, in military dominance, and in the superiority of the Aryan race. Still people killed and gave their lives believing in the superiority of the third Reich proclaimed by Hitler.

The suicide bombers, as well as other people who are indoctrinated (or determined) by their religion or political system and who commit evil acts, act voluntarily but are they entirely responsible for their acts? They wanted to act that way even though they wanted it for reasons which are hard to understand for outsiders.[12] However, to come to terms with the degree of responsibility involved, every single case would need to be investigated. The degrees of responsibility lie on a scale between not being responsible on the one side (as in pure accidental moral evil acts) and entirely responsible on the other side (as in active and responsible moral evil acts).

By *Active but not entirely-responsible moral evil* I intend, for example, immoral actions of people who suffer from neurological or psychological disorders but who had the opportunity to learn what behavior is not acceptable. Their deeds, however, would most probably be considered to have been voluntarily performed. For instance, the manifesto of the perpetrator in Finland showed that the act was indeed intentional. However, the degrees of responsibility, I believe, lie on the scale between belief-based accidental moral evil and active and responsible moral evil.

A candidate for an *active and responsible moral evil* act is physical and psychological torture. Sam Harris condemns torture performed by the Spanish Inquisition because, among other reasons, torture forced people to confess things they never did. Nevertheless he defends torture performed by American officials in Guantanamo.

[11] http://www.infoplease.com/spot/jonestown1.html accessed 15 February 2016.

[12] We are not discussing the sect-leaders, only the members.

He writes, "If there is even one chance in a million that he [Khalid Sheik Mohammed] will tell us something under torture that will lead to the further dismantling of Al Qaeda, it seems that we should use every means at our disposal to get him talking" (Harris 2006: 198). According to him, violence is often an ethical necessity (Harris 2006: 199). Hence, torture performed by American officials serves a good cause (i.e. the intention of the American officials is good) while the intentions of the fifteenth century clerics were not (even though their intentions were sometimes understood as wanting to save a human being's soul). It is needless to say that I completely disagree with torture for whatever reasons. The method of voluntarily torturing people causes unnecessary suffering and pain. Intentionally (voluntarily) causing psychological and/or physical pain or damage to others, whether it concerns human beings, animals or nature, is undoubtedly an act of active and responsible moral evil. People performing such evil acts are doubtlessly entirely responsible for their actions. This reminds me of Hannah Arendt's (1906–1975) words after she attended Aldolf Eichmann's trial in Jerusalem. She said that he was not a psychopath but "thoughtless and terrifyingly normal" (Arendt 2006 in Jones 2014).[13]

We should be careful to relate evil to specific individuals, groups of people or societies, regardless of whether they are religious or not, because to be a human being is to be vulnerable. Human beings are complex beings and morally vulnerable due to their neurobiology, personality, culture and evolution. This makes the problem of moral evil a complex issue. There is no consensus on the commonsense thinking that human beings are genuinely free to act as they will. Rather, it seems to be more plausible that human beings, in some cases, act entirely voluntarily while in other cases they act in accordance with their cultural, psychological and neurological make-up only. Most often, these two cases of action are combined.

Conclusion

On the one hand there are the libertarianists who argue that one is always responsible for one's actions. On the other hand, incompatibilists believe that such cannot be the case due to the fact that humans are dependent on their environment (culture, education, religion and so forth). The problem with the libertarian view is that it needs to be explained in what sense causally undetermined free acts really are free. The problem with the incompatibilist view is that it needs to be explained in what way humans could be held responsible for their acts presupposing everything is predetermined. Libertarianism does not give room for discussion of accidental non-intentional moral evil acts, as suggested by Anderberg. Incompatibilism does not give room for discussing of active intentional moral evil acts as also suggested by Anderberg.

[13] http://www.neh.gov/humanities/2014/marchapril/feature/the-trial-hannah-arendt accessed 21 Aug. 2014.

In trying to contribute to the debate of free will, responsibility and evil, I suggested four types of moral evil. In other words, I tried to broaden Anderberg's distinction. These types were pure accidental moral evil, belief-based moral evil, active but not entirely responsible moral evil and active moral evil. With this it seems that I took a middle road between libertarianism and incompatibilism, arguing that human beings should not always be considered entirely responsible for their deeds, but they should be held responsible in some cases, and sometimes the circumstances are such that they cannot be held responsible at all. However, even when they are fully responsible and are capable of foreseeing the consequences of their deeds, we need to look at the historical, personal (including the neurological make-up) and socio-cultural background of the individuals. Religion is an integral part of worldviews in which non-religious ideologies also have their part.

All worldviews have an individual as well as social (common) dimension. Worldviews, however good they are meant to be, when put into practice can be (and are) sometimes misused. I believe we are in need of a wider philosophical conceptual theory of the problem of evil. I also believe that more interdisciplinary work is needed in order to understand the complexity of the problem.

Today we have access to neuroscientific, nanotechnological, genetic sources, to name some, which together with philosophical ethics might come up with, if not a solution, at least suggestions that could lead towards a solution. Neuroscience is relevant for knowing which parts of the brain are connected to amoral or immoral behavior; genetics is relevant for mapping deviant genes. Also, genetics may contribute with possible solutions by, for example, studying the effects of gene-manipulation. Nanotechnology is relevant as well because it may contribute with possible solutions by, for example, studying the effect of neuro-implants.

However, suppose there are clinical solutions which could minimize moral evil, who would be responsible for the medical files? Who would have access to them? Which social institutions would have the responsibility to determine whether or not and when clinical intervention is necessary? Furthermore, attention needs to be paid to each individual's right to decide whether or not to have clinical intervention.

Finally, what if we are wrong? We have been wrong in the past. Hence, clearly, philosophers are needed to look closely into the ethical and juridical aspects. From the analysis above it becomes obvious that there is evil in the world and I cannot but suggest that evil will not go away. However, the good news is, neither will the good.

Chapter 7
Human Time

Introduction

Another specific issue concerning the relationship between the human being and the world is the notion of time.[1] Hence, the question to solve in this chapter is how it comes about that human beings experience events in time, i.e. that we experience a progression in time or flow of time. What is it that makes us aware, or at least gives us the illusion that there exists some kind of temporal duration that we call time? What is it that lies behind our, illusionary or not, apprehension, feeling or perception that time flows? Julian Barbour writes, "[n]othing is more mysterious and elusive than time" (Barbour 2000: 11). Earlier St. Augustine (354–430) argued that time has no reality; it is only something that humans apprehend. Time cannot be infinite either, because God created it and everything that God created is conditioned.[2] Indeed, human beings 'live' time but cannot really explain it. There are different ways in which people have perceived and still perceive time, of which some are very specific.

In some philosophies, e.g. ancient Hinduism, Buddhism, Pytagoreanism, but also in some ancient cultures like the Maya, the Q'ero Indians of Peru and the Hopi Indians of Arizona, time is perceived as cyclic (cyclic time, the wheel of time or Kalachakra). One could say that time is related to nature itself. The sun rises and sets and seasons follow one after another. Everything comes and goes, people grow old and die but the children reconstitute the process. The way of looking at time is very old but perhaps not so surprising considering how natural phenomena behave.[3]

Some cultures do not have a notion of time at all; they are rare but nevertheless deserve a place amongst different views of time. While the Hopi Indians perceive

---

[1] See further, Runehov 2016a.

[2] Saint Augustine 2011.

[3] On what is time, http://www.exactlywhatistime.com/ancient-philosophy/ accessed 13 January 2016.

A.L.C. Runehov, *The Human Being, the World and God*,
DOI 10.1007/978-3-319-44392-8_7

time in terms of a wheel, they only have two words connected to time, sooner and later. A culture that does not have any time orientation is the Pirahã tribe of the Amazon rainforest. They have an language lacking numbers and words for colors and time. They do not have art, religious beliefs or creation myths. They literally live in the present "when they can no longer perceive something, it effectively ceases to exist for them" (Ancient-Philosophy).[4]

However, the most common way of perceiving time is experiencing it as a flow, or a march. This way of thinking about time is linear. What is it that makes us aware, or at least gives us the impression that there exists some kind of temporal duration that we call time? What is it that lies behind our apprehension, feeling or perception that time flows? Human beings seem to perceive time as moving from past to present to future, or, from future to present to past. For example, suppose you are sitting in a train you might wonder whether it is time that moves toward you, or is it you moving through time?

Psychologically speaking, humans respond to their environment in such a way that they experience one perceived event preceding or succeeding another. Henri Bergson (1859–1941) distinguished between time humans live by and mechanical time as seen by the sciences. In other words, one can look upon time as *experienced* and as *studied.* Experienced time is dynamic, studied time is static. Humans perceive time through a succession of separate, discrete, spatial constructs, more or less like watching a movie. However, there are different opinions of how the flow of time can be experienced.

Bergson argues that humans experience time as a continuous flow, without a beginning or end ([1889], (1965)). According to St. Augustine, the experience of the flow of time is rather a matter of *transience* (a kind of moving on); for others it is a matter of *succession* (one event after the other); and yet others argue that the human brain has some clock-like device. Physicists argue that our perception of time can only be examined in relation to perception of other events. This line of thinking turns the experience of the flow of time into a function of psychological and physiological processes only. As marked upon by Barbour, actually physicists do not ask the question what time is and leave solving the riddle of time for philosophers. There seem to be different explanations of the notion of time which are not always in agreement. How humans experience time is not the same as how scientists explain time. Furthermore, scientists have explained the notion of time in different manners. In order to show these discrepancies I will discuss time as it is experienced and studied. The scientific views of time presented are Newtonian time, Einsteinian and Quantum time.

Experiencing Time

Why do we experience time as we do? Several explanations have been given.

One of them is known as the problem of the specious present. The specious present, or conscious or mental present, is generally measured by determining the period

[4] http://www.exactlywhatistime.com/ancient-philosophy/ accessed 13 January 2016.

of time over which stimuli may be spread and yet conceived as one temporal unit (Frankenhauser 1959). The most investigated example for the conscious present is music. How is music perceived? When we listen to a melody, we perceive groups of notes, woven together, stretched in time, but still we do not perceive them simultaneously. (If they were experienced simultaneously the sound perceived would be a chord, which is something completely different). Listening to music is an auditory experience in which we are not directly aware of separate temporal events. Nevertheless, we seem to experience some kind of time flow. What could this be due to? Several propositions have been submitted.

One is that memory is the cause of the illusionary perception of time progression. Fortunately or unfortunately, this proposition fails. Indeed, it seems peculiar that while listening to a piece of music we should be aware of hearing the note *Mi*, whilst having a short-term memory of having heard the note *Do* and a short-term memory of having heard the note *Re*, whilst having a short-term memory of having heard the note *Do*. This example concerns only three notes; imagine listening to one of Bach's Fugues! That would imply that we would have memories of memories of memories of memories of .... in a huge Russian nested-dolls (matryoshka) style. Lockwood writes, "such nesting of short-term memories within short-term memories like Russian dolls is a psychological absurdity" (Lockwood 1989: 266).

Another proposal is Franz Brentano's phenomenological solution (2015). Even though this proposal is a better one than the memory one, it also fails to explain our time perception in a satisfactory manner. Brentano's thought is that one and the same phenomenon can be presented to our consciousness in different ways, and this is likewise true with things we are not immediately conscious about. For instance, the note *Do* can present as *present Do*, *recently past Do,* or *further past Do*. Imagine hearing the scale Do Re Mi. The note Do will present itself as the present note, quickly followed by Re presenting itself as the present note together with the note Do just past. This perception will be followed by Mi, presenting itself as the present note together with the note Re just past and the note Do further past. Just past and further past are two representative positions on a phenomenally continuous spectrum of temporal modes.

The problem here is that we simply do not perceive in such a way; we have no need to analyze how we perceive; we just do. We perceive a piece of music as it presents itself to our consciousness in a temporal order. Another problem with this proposal is that 'just past' could be interpreted as an act of memory similar to Bertrand Russell's (1915) concept of *immediate* memory, meaning that "we are, in his technical sense of direct awareness, *acquainted* with past sounds, shapes and so forth that lie within the specious present" (Lockwood 1989: 269).

Yet another proposal is that we confuse phenomenal change with temporal passage. The problem is that there is nothing inherently dynamic about change. Change merely means that different times are associated with different phenomenal attributes. Lockwood, inspired by David Hume's distinction between first and second order causation (2008) divides the concept of change into a first order change and a second order change. A first order change is a change in tone, such as when jet fighters fly overhead. This change is analogous to the variations in color within the visual field when looking at the sky. A second order change has no spatiovisual analogue. By the second order change Lockwood means variations in the way that a single phenomenal item is presented from one phenomenal perspective to another.

According to Lockwood, if we compare this idea of dividing the concept of change into a first and second order change with Hume's theory of causation, we will see that it does for temporal flow what causation does for the concept of causal connection. Hume said that there is nothing more to causation than a constant conjunction of similar pairs of events. When on several occasions we experience event A closely followed by event B without ever experiencing A without B, we will expect a B whenever we experience an A. This expectation is then illicitly projected onto reality. We will think that A and B are connected by some relation of necessity (Lockwood 1989: 278). The Humean account is not completely persuasive though. There may be some truth in this, psychologically speaking; however, the fundamental reason why we experience things the way we do is simply because things are that way. There are simply causal connections in the world and we do, in some way, progress through time.

However, the experience of time is sometimes (if not always) relative to the experiencer. Sometimes one experiences time as moving very slowly; for example, when one is bored or longs for something; and sometimes one experiences time as moving too fast, for example when one is very busy or on holiday. Sometimes one has the feeling that time has stopped. Hence, some aspects of time are very personal. Perhaps it is the case that people possess a kind of psychological clock. On the other hand, the experience of time is also intersubjective. I believe that this aspect of time can be understood as a deontic power, because this aspect of time is created by us.

Indeed, in order to have an international civil time standard, Greenwich Mean Time (GMT) was created, which was later superseded by the Coordinated Universal Time (UTC). I suggest that it is the deontic type of time that is not shared by the Pirahä tribe, because they have not yet conceptualized time in such a manner. Actually they have not yet conceptualized time at all due to their extremely limited language. On the other hand, since all people have a very similar biology, also they would experience time, only not as expressed. Generally one could argue that, even though humans experience time in a personal manner, it does not imply that time is relative, thanks to human creativity and deontic powers. It is time to look at time as studied.

Newton Time

It was Isaac Newton (1643–1727), who put forward a theory of precise notions of space, time and motion (from 1687). He understood time and space in a common-sense way. Newton's time is absolute. Nothing can disturb the flow of time. Also space is absolute and understood as a "limitless container" (Barbour 2000: 20). Space is seen as a "huge arena; time [as] a clock in the grandstand" (Barbour 2000: 20). The Newtonian view prescribes a three-dimensional reality that moves forward

Fig. 7.1 Newton time in terms of an arrow

in time. In this temporal order there is a past, a present and a future. Time itself is absolute and moves constantly without being affected by external features. However, just because Newtonian physics follows a commonsense understanding, this does not mean that it is without philosophical problems. For example, Newton's physics introduces a deterministic understanding of the universe and everything in it. "At any instant, all the things in the Newton world are at definite positions" (Barbour 2000: 20). Everything in the past, present and future is fixed. Indeed, with the help of Newton's laws, based on the mutual effect between particles, it is possible to precisely predict the behavior of a particle. To put it differently, if it were possible to know both the location and velocity of every particle in the universe, one could *know* the future of the universe. That would imply that every human being's life is determined. Free will becomes an illusion. The flow of time, which is fixed, is often presented as an arrow moving from left to right, representing an infinite duration of time, in which the left side represents the past and the right side the future. Somewhere along the arrow between the left and right is the present, which is often marked by a dot. Fig. 7.1.

Needless to say, such an understanding of time raises philosophical concerns.

For instance, suppose we can imagine time as an arrow, what would be the past's duration, present and future? Is the present simply a dot? Or does it extend towards the past and the future, and if so how far? Clearly our experience of the present includes moments of both the past and the future in terms of memory and in terms of plans and dreams and expectations. Exactly where on the arrow would we put these memories and plans? Newton's time is a fixed linear one, comprehended as a line with a direction from the past to the future.

Ludwig Boltzmann (1844–1906) argued that there is no flow of time, no direction. At best, time is a line. "It is only the instantaneous configurations of matter, strung out like washing on the line, that very occasionally suggest that time has a direction associated with it. The direction is in the washing, not the line" (Boltzmann 1895 in Barbour 2000: 27–28). Furthermore, Boltzmann argues that depending on the position on the line, the arrow points in opposite directions. Boltzmann still sees time as a line but one without directions. Time is not like an arrow, there are several possible arrows on the time line.

Boltzmann's idea is more fruitful than Newton's in the sense that time is understood as a dynamic phenomenon. For example, if I plan to attend a conference in some months, I will pay my registration fee, book a room, buy travel tickets, and so forth. This means that 'time' *in my mind* is heading for the future. But if I write down some memories of past events, time *in my mind* returns to former episodes in my life. Hence, is time merely a psychological phenomenon after all? Especially when realizing that my planning the conference in the future as well as collecting some memories in my diary are done in the moment 'now', which mentally comprises future in the former case and past in the latter. Is time then illusion?

Einsteinian Time

Albert Einstein once said that time and space are modes by which we think and not conditions in which we live. Indeed, things changed when Einstein entered the arena of physics. With Einstein, time became a fourth dimension in addition to the three dimensions of space. Space and time were no longer separate. What happens in time also happens in space and vice versa. His way of thinking implied that reality became a hypercube in which everything has its place, the past as well as the future, and everything in between. In other words, when a group of people are in the same room they are in the same position in space and time. They are in the same position in space as the people who were in this room previously, but obviously not on the same time line.

Einstein introduced two new concepts; *coordinate time* and *proper time. Coordinate time* implies that there is no universal march or flow of time. This is so because there is no universal present; and therefore no universal past or future. According to Einstein's Theory of Relativity, attributions of length, or of temporal separation between events, only make sense when they are understood as relative to a chosen frame of reference. For example, suppose two persons move towards each other. Since they are moving relative to each other (due to having different frames of reference) they will disagree about the temporal and spatial separation between the two events taken individually. However, they would agree upon the spatiotemporal interval, which would be the result of a measurement of spatial distances in three dimensions, with time as the fourth dimension. The result of this measurement, known as *space-time distance (or proper time)*, turns out to be the same for all observers, regardless of their state of motion (Flood and Lockwood 1986: 6–9).

Nevertheless, neither coordinate time nor proper time involves any notion of time passing, or the flow of time. Consequently, the flow of time, or passing of time, has to be consigned to the mind after all. However, not all has been said yet.

Quantum Time

Erwin Schrödinger argued that the "theory of relativity … however revolutionary leaves untouched the undirectional flow of time, which it presupposes" (Schrödinger 1967: 152). Einstein introduced a new revolutionary theory but as he said,

> to create a new theory is not to destroy an old barn and to build a skyscraper instead. Rather it is to climb a mountain, get broader views; discover unexpected connections between our starting place and its comprehensive surroundings. But our starting place still exists and it is still visible only it has become smaller and builds a little part in the wide view which we achieved through conquering the obstacles on our adventurous way to the top of the mountain (Zukav 1979: 45).[5]

[5] My translation from Swedish.

Quantum physicists such as Niels Bohr (1885–1962), Werner Heisenberg (1901–1976), Erwin Schrödinger (1887–1961) and others further explored the mountain.[6] They thought along completely new lines something which at the beginning of their physical enterprise was seen a strange and mystical. This let Einstein to utter his famous words: "It is hard to sneak at God's cards. But that he would choose to play dice with the world... is something I cannot believe for a single moment" (in Calaprice 2005: 237). Indeed, and in order to understand quantum time, we need to consider the measurement problem of quantum physics; i.e. we need to understand:

> [...]what, in quantum [physical] terms, happens when we observe a physical system. [...]. What the quantum-[physical] measurement problem is really alerting us to, [it] is a deep problem as to how consciousness (specifically the consciousness of the observer) fits into or maps on to the physical world (Lockwood 1989: preface, p.x.).

Newton had a dualistic understanding of the mental and the physical. In Newton's world, everything has its given place. Einstein changed the understanding of space and time. Nevertheless, the mental and the physical remained separate. This changed with the introduction of quantum physics. From now on, the mental and the physical receive the same status and are seen as equally problematic. Hence, as already mentioned in the chapter on the self, mental states and events follow the same temporal and spatial order as physical states and events (Lockwood 1989: 75).

The Measurement Problem

The measurement problem of quantum physics deals with the problem of *where* to *put* the border between the *classical world* and *the quantum world*; i.e. between the macro world and the micro world. *The border problem* is also known as the problem of the collapse; i.e. when does the wave function collapse into specific values?[7] In other words, the question is what happens when we observe a physical system?

To repeat what has been mentioned in the chapter on understanding reality, the basic physical variable used in quantum physics is the wave function, Ψ, as described by the Schrödinger equation, which states that if we know Ψ at time zero and we know the Hamiltonian, then we can calculate Ψ for all future times. The first postulate of quantum mechanics is that all information about a physical system is to be

[6] Bohr, Heisenberg and Schrödinger and others, see Bohm and Hiley 1993. Recommended for extended view on the different interpretations of quantum physics and its history.

[7] It is important to note that the concept of collapse was mostly used by von Neumann, *"Mathematical Foundations of Quantum Mechanics"*, Princeton, 1955. Neither Bohr nor Heisenberg used the term collapse. Von Neumann had a more realistic view on the wave function; he was concerned with criticism of the measurement problem. Bohr and Heisenberg avoided the measurement problem by having a non-realistic view of the wave function which made it possible to avoid the collapse. It is important to notice the distinction between the von Neumann view and the Bohr/Heisenberg view.

found in the wave function, Ψ of that system. If we manipulate the wave function Ψ[8] which we do when we measure or observe a particle, then we receive information about the location, energy and velocity of that particle. The wave function is interpreted in terms of probability, which means that we can only specify (for example) a probable location of the particle. Let us take dice as an example, given by Henrik Carlsson.

Quantum physically speaking, we could understand a die in terms of a superposition of all its six possible states, 1, 2, 3, 4, 5, and 6. Once the die is thrown of course only one state will be on top, 1 or 2 or 3 etc.... In quantum physics, the moment the die is observed on the table, the wave function Ψ of the die-system collapses into one of its classical states 1 or 2 or 3, etc.....[9] The basic attribute or characteristic of quantum physics is that whenever we measure or observe a physical system, the system collapses to one of its possible classical states. We do not know which state or states the system was in before measurement or observation took place. Not knowing which state or states the system is in before measurement is, so I believe, what Bohr meant with his famous words that *nothing exists until it is measured*, (i.e. no photon exists until a detector fires).[10]

Is there something problematic about the collapse? As mentioned above, when we measure a physical system (quantum-physically speaking), the classical world (the measuring apparatus and the observer) meets the quantum world (photons, waves, particles). This meeting causes the system to collapse. The problem, it seems, is to know *when* the collapse takes place.[11] According to Eugene Wigner, the collapse occurs when the observer becomes *aware* (conscious) of the result of the measurement. He argues, "[...] the impression which one gains at an interaction, called *also the result of an observation* modifies the wave function of the system. The modified wave function is, furthermore, in general unpredictable before the impression gained at the interaction has entered our consciousness. [...] It is at this point that the consciousness enters the theory unavoidably and unalterably" (Wigner 1967: 172–173).

Hence to Wigner, it is not enough to measure a system; one has to be aware of the measurement as well. One consequence of the collapse is that, as Lockwood writes, as soon as an observation is made and the undiminished reality (wave function) breaks down (collapses) possible comprehension of that undiminished reality is lost. As observers, we need to disturb the system when we observe it. We cannot conceive of what the wave function is in itself. We can only conceive the *eigenvalues* (one of the possible outcomes) resulting from a performed measurement or observation on the wave function.

[8] It might be interesting to note that Schrödinger gave the wave function Ψ an ontological meaning; he meant that the wave function Ψ exists independently of our knowledge about it.

[9] The example with the die derives from a lecture given by Dr. Henrik Carlsson at the department of Quantum Chemistry, Uppsala University. The literature behind is *Kvantmekaniken, ett försök att beskriva det allra minsta,* Carlsson, H. 19xx: 17–21.

[10] Bohr, Niels, http://www.newworldencyclopedia.org/entry/Niels_Bohr

[11] This is perhaps why there are that many interpretations.

In my view, this confirms what Immanuel Kant held many years ago; i.e. that we cannot have knowledge about the *thing* as it is in itself (Kant 2004). It also confirms Bertrand Russell's argument that we can only think of matter *by description:* as that which stands in a certain relation to our perception (Russell 1915, in Egner and Denonn 2009). Let us return to the topic of this chapter and ask how time is understood within quantum physics.

Quantum physically, time is problematic and not exact. A general interpretation of time is that it is a parameter; some kind of label, which marks a certain state. In this sense all times are equally *possible.* To repeat, before observing a physical system, the system is in a superposition of all its possible states. This superposition then also includes all possible times. Suppose we want to measure a particular physical system at 3 pm; then the probability of *getting* 3 pm is greater than getting 5 pm. However, since time is but a label, where does this leave us? The problem is that 3 pm is a classical physical time. There can only be one time quantum physically: the present. Past and future only exist in the present. This means that, quantum physically, we can relive past events, but only in the present; we can imagine future events, but only in the present, something which I mentioned above, and which to me makes completely sense. As such we are free to live the past and future at will.

This seems to imply though – which also makes complete sense – that we can change the events of our past and build events in our future. For example, I can imagine that I received this wonderful pink princess dress from Santa Claus when I was a child, a dress I actually never got. I can live myself into this particular event knowing that it is not true but as if it were true.

Time hence becomes a brain observable in its own right (Lockwood 1989: 283). Time is regarded as an element in a preferred set of brain observables. In this way, like all the other brain observables e.g. colors, time becomes an element of a superposition of the *preferred* set of brain observables, be them introspective or factual. Lockwood writes,

> Suppose it were possible to construe time itself as an observable in the quantum-mechanical sense. As such it would have a spectrum of eigenvalues that corresponded to a range of possible times. And each of these different times would appear as the time – as now, that is – from the standpoint of a corresponding phenomenal perspective (Lockwood 1989: 283).

To Lockwood, if the sense of the flow of time, the sense of progression through time, is not an illusion, it surely is but a perspective-relative phenomenon, something that occurs only from the point of view of the sentient being itself. The fact that we use external clocks does not change the states of things in themselves, save for the relationship between their states and the states of ourselves.

A given phenomenal perspective corresponds to a set of eigenvalues of some preferred set of compatible brain observables. Only if time observables are themselves included in the set associated with phenomenal perspectives can these perspectives collectively come to generate streams of consciousness. However, what is intuitively true is that we are simply aware of some kind of time passing. We cannot explain how we know that we are aware of a passing of time, but we are aware of ourselves being present and of others becoming present in our present. We are aware of things and phenomena becoming present in our present. We are aware of our experiences and experiments. We know that, as Kant put it, whatever we experience, we experience within a space-time frame (Kant 2004). We are also aware of a possible tomorrow that will possibly become our next present; and that yesterday can *never* become a new present as a memory in the way it was present yesterday. The past enters the present as a correct or constructed memory, the future enters the present as a wish or planned event. But we are aware of that one-way time ticket from physical birth to physical death, a ticket valid for all of us and everything around us.

It is true of course that when we observe something, we observe it at time 'now'. There is no other possibility. If we want to have access to a previous time, we have to recall the *eigenvalues* of a previous observation; but this previous observed time is also observed as now. It is within our own biography that we build a history, as it is only within our biography that the *eigenvalues* of our different observations, including all the *nows* of the different observations, are to be found. By way of memory we have access to our history. It is within our own history, our own biography that we mistakenly believe in experiencing a passing of time.

May it be due to the way in which our sensory responses are constructed? It is within our own biography that we know *what it is like to witness a passing of time*. But as we can only answer this question for ourselves, the passing of time seems to be a subjective matter. This, I believe, is in line with what I said above, namely that except for the noetic power of time, time is a personal thing.

Conclusion

It seems that all apprehensions of the phenomenon of time have something to say about how human beings experience time. Humans do experience a flow of time, even though a flow of time cannot scientifically be proved. Still, in my opinion, how time is understood in quantum physics becomes the closest to how humans experience time. In other words, humans experience all time from within the present. We do not experience one-way-direction from the past to the future.

Human consciousness is constructed in a manner that it can mentally live future events, and relive or reconstruct past events, but, only from the present. This leaves open the capacity of creativity, something which is not a bad thing for oneself, one's environment and one's survival. Indeed, sometimes, having the capacity and psychological strength to reconstruct past events can save one's life, not least psychologically. Similarly so can having the capacity and psychological strength to look positively at the future.

Conclusion Part II

The second part of this book was concerned with how human beings relate to the world. What needed to be explained was how external reality is apprehended. While internal reality is sometimes mentioned as well, this was discussed in the first part.

Discussing external reality brought me to the philosophical debate of realism which concerns human capacity to describe reality in a reliable manner. Several types of realism were investigated.

Metaphysical realism seemed to be adequate only when it comes to describing artifacts; because human beings are not part of the artifacts, they create them. However, when it comes to describing chemical, physical or quantum reality, metaphysical realism does not prevail. This is because human beings are part of that type of reality, and to be able to say that such reality is as humans perceive it would imply having a view of nowhere, having a God's eye view. This was shown by way of quantum physical descriptions. When trying to describe (explain) one particular quantum reality, different interpretations prevail.

Anti-realism was shown to be adequate for describing some internal or subjective realities but does not succeed in adequately describing intersubjective reality, reality shared by us.

Internal reality then, which argues that we cannot express correctly what we cannot conceptualize without therefore intending that there is nothing more to say about reality was shown to be the most fruitful candidate of the three.

However, it has its limits. This is why I suggested a new type of realism, namely extended realism, which includes, firstly, measurable or observable reality (e.g. scientific measurable reality). This is reality *for us.* Secondly, it includes creational reality (e.g. artifacts), which is reality *from us for us*. And lastly, phenomenological reality (e.g. colors), which is reality *within us for us.*

The next subject matter for discussion concerned the philosophical debate of naturalism. The first one, comprehensive ontological naturalism was shown to be not fruitful at all. It is not the case that all x are (always) y.

Two new types of naturalism were introduced. Extended or flexible ontological naturalism, which says x or y and minimalistic ontological naturalism which says that it is only within a specific realm R that all x are y. These types were shown to be much more fruitful to adequately discuss reality. I ended the discussion on realism and naturalism by comparing the two philosophical strategies to describe reality. Methodological and epistemic naturalism seem to be linked to internal or pragmatic realism because of the focus on methods, be they scientific or epistemic. Internal or pragmatic realism also seems to have some things in common with minimalistic ontological naturalism, not least that reality is seen within a realm. Extended realism and extended ontological naturalism are the types that can most closely be linked together.

Human being and the world also concerns human actions in the world. Hence, the subject matter of mindreading, empathy, compassion, free will, responsibility and evil needed to be investigated. Mindreading concerns what we mentally do when we interact with other people, as the term says; we read other people's mind. However, we also read our own minds and in addition, we write our own minds. Mindwriting is however only possible on the first-person level. Hence, humans are mindreaders and mindwriters in relation to the first-person perspective, and mindreaders in relation to the third-person perspective.

The core of mindreading, it was argued is empathy. I divided empathy into low – and high level empathy. Low level empathy comprises emotional contagion and biological altruism. These types of empathy are emotional, immediate and involuntary. High level empathy comprises sympathy, selfless altruism and agape which are both affective and cognitive and linked to consciousness. The bridge between the two levels, it was said, is compassion. Philosophically, empathy (high level empathy) is established by way of analogical inference. This view has been questioned, not least by the mirror neuron theorists who argue that there is no need for theorizing in order for empathy to occur. I argued that this may be the case, but only when it concerns low level empathy, since this type of empathy is only affective and involuntary.

To show my point I divided the mirror neuron view into the closed mirror neuron view and the open mirror neuron view. I did this on the basis of neurological conclusions drawn from studies on empathy. I argued that the closed mirror neuron view is applicable to low level empathy because these mirror neuron mechanisms are restricted to the F5 area, and are motor processes which are automatic and immediate. The F5 mechanisms concern representative mirroring others' actions and behavior. In other words, monkey see, monkey do, there are no questions asked. The open mirror neuron view gives room for higher levels of empathy because it concerns imitative mirroring, includes a process of analysis, and cannot be established by action-mirroring alone. Neurologically there are more neurofunctions involved than the F5 area. That high level empathy correlates with different brain structures, not least the cognitive structures, has been shown in various neuroscientific studies.

Compassion occupies an important place on the mindreading scale. The question asked is whether compassion is a good or not so good thing. Two views of compassion were investigated, the anti-compassion view and the pro-compassion view. The first view has it that all people are equal and that compassion disturbs this equality, because it takes away the dignity of both the compassioned and the compassionate. Furthermore, one should never trust one's emotions and compassion is an emotional thing. Hence, showing compassion is not good at all.

The pro compassion view does not dismiss the affective part of compassion but still restricts it. Therefore this view suffers from several problems; the problem of misplaced compassion; the problem of blameless suffering, of social and ethnic diversity, the problem of other species and of undeserved sufficiency. Compassion is both affective and cognitive but it is the cognitive part that should be emphasized. A problem of compassion, regardless of view, is that it is unclear why people act compassionately. Do they do this for the sake of the other person, for the sake of their own self, or for the sake of both?

The questions raised in the chapter on free will, responsibility and evil were whether there is free will and if so, to what extent can people be taken to be responsible for their actions. Different types of evil were presented and analyzed.

I took the middle road between libertarianism and incompatibilism, arguing that human beings should not always be considered entirely responsible for their deeds; sometimes they should be held completely responsible and sometimes the circumstances are such that they cannot be held responsible at all, e.g. due to coercive psychological manipulation or a dysfunctional neurological make-up. In order to show that, I broadened Anderberg's distinction of moral evil becoming pure accidental moral evil, which does not demand responsibility; belief-based moral evil; and active but not entirely responsible moral evil; and active moral evil, which demands full responsibility. It was argued that it is important to always consider the historical, personal (including the neurological make-up) and socio-cultural, including religious background for the individuals' behavior.

The question about how to understand time was also a question that needed attention since human beings have experiences of time, even though the idea that there is something like time, a flow of time, cannot scientifically be confirmed.

Humans experience time in different ways, but the most common is the experience that there is a time direction from the past to the future. Newtonian time is in line with this commonsense thinking. Newton's time is fixed, presented as an arrow moving from the left (the past) to right (the future) representing an infinite duration of time. To Newton, time and space were separate. This changed with Einstein who established a four-dimensional structure where space occupied three dimensions and time one. With this came two new concepts of time, coordinate time and proper time. None of these include any notion of time passing. With quantum mechanics, time became a label, time only occurs when a measurement or observation takes place. This means that there is only one 'time' which is the present, the now, the moment of the observation. I argued that the quantum physical way of understanding time comes the closed to way in which humans experience time.

With this I meant that, after all, even if humans refer to a flow of time, they can only refer to it from the present, from the now.

Part III
God

I am a little pencil
In the hand of a writing God,
Who is sending a love letter
To the world.

Mother Teresa

Introduction

We have now come to the last part of the present manuscript which investigates some ideas on the relationship between human being (esse and ens) and God or ultimate reality.

A first question that might be asked is whether all people have a relationship with God or ultimate reality. My answer to this is yes; even atheists relate to God, in the sense of *not God.* Atheism is however understood in different ways. Hence, a brief presentation of different understandings of atheism is needed. Another important question is how to understand God or ultimate reality? Once again, there does not seem to be a consensus about how to understand the divine. Indeed, Jeanine Diller and Asa Kasher required 1041 pages in order to account for different interpretations of God or ultimate reality (2013). Obviously, I will only be able to present some of them. However, these models are important for the main discussion of how God or ultimate reality could act in the world. How God or ultimate reality could act in the world is one of the main topics of debate not only in the philosophy of religion but also in the field of science and religion. Once again there is no consensus. Hence, once again several ideas are discussed which result in my own idea of how divine action in the world could be understood. The model I suggest is mainly, but not only, inspired by Arthur Peacocke's (2002 and 2004) and Philip Clayton's (2004) and

Nicolaus Cusanus' (in Sandbeck, L. 2007) panentheism. Combining those ideas with Linda Zagzebski's view of God having personhood and Celia Deane-Drummond's idea a God-human relationship in terms of friendship, gave me the opportunity to establish a model of divine action in the world that also included the idea of emergence as well as the idea of the *imago Dei.*

Chapter 8
God-Human-God Relationship

Introduction

The question is how God or ultimate reality could be understood. Is it even possible to grasp or model God or ultimate reality? In *Models of God and Alternative Ultimate Reality* (Diller and Kasher 2013) Robert Cummings Neville hypothesizes that "ultimate reality is an ontological act of creation, the terminus of which is everything determinate, constituting and unfolding in spacetime" (2013: 19). Defined as such, he argues, ultimate reality cannot be modeled and anything that can be modeled cannot be ultimate reality. The question is, if we accept this definition, how could human beings and ultimate reality or God interact? Or, in the words of Philip Clayton and Arthur Peacocke (1924–2006), "[h]ow, in the light of the sciences, to conceive of God's relation to the world as it is now perceived to be and to be becoming" (2004). Hence, in this chapter some models of God or ultimate reality will be presented and discussed; I will also explore how a relationship between humans and God or ultimate reality could be understood. Furthermore, philosophical problems such as the problem of time and of free will in relation to ultimate reality will be investigated.

Another question that is raised is whether everyone relates to God or ultimate reality. Some would argue that this is not the case; for example, atheists do not relate to God or ultimate reality. However, in my opinion they do, because arguing that there is no God or ultimate reality is actually taking a stance, and taking a stance is a way of relating to something. For instance, in order to *raise* people's *consciousness*, Richard Dawkins needs to put God in a bad light (2006). Hence, God becomes some kind of supernatural person possessing all the negative characteristics found in the scriptures and other religious and historical writings. Alister McGrath made a brief summary of Dawkins' view of God: God is "a petty, unjust, unforgiving control freak; a vindictive, bloodthirsty ethnic cleanser; a misogynistic, homophobic, racist, infanticide, genocidal, filicidal, pestilential, megalomaniacal, sadomasochistic, capriciously malevolent bully" (2007: 46).

A.L.C. Runehov, *The Human Being, the World and God*,
DOI 10.1007/978-3-319-44392-8_8

Clearly, Dawkins made a great effort in his relation to God. In short, religious, agnostic, and atheistic people do relate to God or ultimate reality in their own way. Hence, I need to say something about atheism, without digging too deeply into the subject matter.

Models of Atheism

It has been brought to my attention that types of atheism might be seen as models of God and vice versa. I agree with that. For instance, Ted Peters chose to include atheism within his models of God. (2013: 49–51). However, that I chose to separate types of atheism and models of God is, as mentioned above, that I do not wish to dig too deeply into the subject matter, but felt that some clarification is needed.

Atheism is commonly understood as believing that there is no God or ultimate reality; that there is no life after death; that everything comes and goes in a natural way, from nature and back to nature. However, historically (Western) atheists were seen as those who did not confess to the *true religion*. Atheism was equated with paganism and also with treason. In 399 BCE the Greek philosopher Socrates was sentenced to death because the jury, consisting of 500 Athenians, accused him of refusing to recognize the gods recognized by the state. History is filled with such examples; e.g. the persecutions of Catholics in England, Huguenots in France, Protestants in Belgium, Germany and the Netherlands. Having the wrong religion was equal to being a heretic or an atheist in the sense of being *atheos*, i.e. godless.

Today atheism is understood in different ways and it is not always clear what is meant by it. One view is that those professing atheism do not believe in a personal God. Hence, some (but not all) Eastern religions are considered to be atheistic in that sense, for example Buddhism.[1] Also Gnostics are sometimes referred to as atheists because their view is that the way to salvation for a human being is to complete oneself by way of attending to the introspective spiritual self. Gnostics regard human beings as embodied spirits. Gnosis (Greek, gnosis) means knowledge or insight. This knowledge or insight can be arrived at by way of interior, intuitive means only. Gnosticism is an intellectual endeavor asking questions about who we are as human beings, where we come from, and where we are heading, historically and spiritually (Jonas 2001). Even though these questions are philosophical in scope, the answers given by the Gnostics are more a kind of mysticism than philosophy.[2] The body was considered to be negligible; only the spirit was important to the Gnostics. For instance, Basilides of Alexandria (117–138 AD), a Gnostic philosopher, argued that the death on the cross of Jesus Christ was only apparent; it did not

[1] Indian religions are polytheistic for example.

[2] Most Gnostic texts date from the early Christian years, but it should be noted that Gnosticism arose before the arrival of the Christ.

actually occur *in the flesh*.[3] For Christians, the death on the cross of the Christ is a crucial doctrine.

Deists are yet another group of religious people who are sometimes considered as atheists even though deists believe in God. Initially, deism was a movement or system advocating natural religion and emphasizing morality. However, since the enlightenment period, any interference of a creator with the laws of the universe is denied. Since such interference is typical for theism, deism is ranked among atheistic views. Agnostics, who do not confess to any ultimate reality and who would not call themselves spiritual but do not deny that there might be some transcendent factor in the universe, e.g. energy, are sometimes also referred to as atheists.

However, most commonly, atheism is understood as the view that there is no such thing as a God and no such reality as ultimate reality in any sense. All there is, is part of the circle of nature; i.e. things come and go, including human beings. Once gone, there is nothing left. Among those adhering to this view, there are those representing a stronger version, which is referred to as fundamentalist atheism. One of the elements of this fundamentalism is that they understand themselves as having the task or mission of convincing other people of their own view, e.g. Richard Dawkins (2006). Because of his mission, some have even called him a religious atheist. While Dawkins thinks badly about the monotheistic religions, he does recognize Buddhism, however, not as a religion (which it is) but as an ethical or philosophical system. Perhaps he confused Confucianism, which is indeed an ethical philosophy, with Buddhism. Even though there is far more to say about atheism and all its variants, I do not see it as my task to engage in this debate at this point. Neither will I dig into the subject matter of religion or religions. My question is rather, how, in the light of science, can we understand a relationship between God or ultimate reality, human beings and their world? We need to look at how God or ultimate reality has been understood or modeled.

Models of God

The problem with a human-God relationship is that such a relationship rests on a personal basis and is hence a subjective matter. Relationships are both cognitive and affective in scope. Though there may be, and are, intersubjective experiences of ultimate reality (see chapter two above), whether there is a continuing relationship or not, as well as what kind of relationship is established, remains entirely personal. Perhaps this is why it is claimed that God is ineffable, e.g. Rudolf Otto (1869–1937). For instance, Otto explains a religious experience in terms of its awesomeness, overpoweringness, of energy and urgency, of the Wholly Other and of Fascination (1970: 13, 19, 23, 25, 31). Observe that I am not talking about religions or faith or belief, which is intersubjective through its institutions. I am talking about

[3] More, E., http://www.iep.utm.edu/gnostic/print accessed 18th January 2016.

a relationship between God, the world and human beings. How such a relationship is apprehended remains personal.

Nevertheless, people have tried to model God and hence several models of God are available. Perhaps the most classical model of God is the Christian one in which God is modeled as a father figure. After all, Jesus' prayer begins with the word, 'our Father in Heaven'. However, since the middle Ages, God as a father figure reigned over the world from his throne in heaven, judging the good and the wicked. Justin Barrett has a theory claiming that all children (at a young age) naturally "assume that [there are] agents [who possess] divine properties" (2004: 108). Such properties are superpowers, super-perceiving, super-knowing, immortality etc. Even though these beliefs come naturally, they are not well developed. God is seen as a super-person, a superman. This seems to indicate that all people, at least in their early childhood, have a similar God-belief. Children at a young age model God as a person, a super-person. Barrett's theory is supported by Ernest Harms (1944); Hermann Siegenthaler (1980) and Helmut Hanish (1996). These studies were analyzed by Newberg and Waldman (2009). In all studies performed in Germany and the USA, children were asked to draw a picture of how they imagined God to be. Interestingly, children below the age of six pictured God as a human being but attributed to God a higher position (e.g. king); older children replaced the image of God as a super-person with symbolic images (e.g. a cross) and the oldest children's drawings were even more abstract (e.g. sun rays). One older child returned his paper in the same way he received it, i.e. blank, meaning that God cannot be pictured. However, 80 % of the non-religious 16-year-olds still drew pictures of God as a person but powerless and weak. (Newberg and Waldman 2009). The latter seems to fit very well with how Dawkins pictures God. Dawkins' idea of God has not developed since early childhood. One philosophical problem is also that 'person', 'personal' and 'personhood' are understood as synonyms, which is not correct, of course. I will return to this below. In the meantime let us look at how God came to be seen as a person.

Time travelling to the era of the Old Testament, David's psalm tells us that "[t]he heavens declare the glory of God; the skies proclaim the work of his hands" (Psalm 19:1). Hence, to think about the heavens is to think about God, and to look at the skies is to look at God's creation. Understanding God in this way is to understand God in terms of a being, as 'ens'. God is seen as the creator of everything that was, is, and ever will be; but could God be an ens in the same sense as a human ens? Some religions (e.g. the monotheistic religions) have personified God, or more accurately, super-personified God. God has become an almighty person. Not all religions have developed an understanding of God along this line; some apprehend ultimate reality as pure consciousness (e.g. some variants of Hinduism and Buddhism), and others have come to see ultimate reality as emergence (e.g. Daoism and Confucianism). (Neville 2013: 19–20). According to Neville, none of these models seem stable and I would agree. Nevertheless they have their advantages.

The advantage of the Abrahamic religions which model God as a person is that God has a name (Yahweh, Allah). God as a person created the world and interacts with the world. In Exodus 3:8 God spoke to Moses: "And I have come down to rescue them [the Israelites] from the clutches of the Egyptians and bring them up out of that country, to a country rich and broad, to a country flowing with milk and honey, to the home of the Canaanites, the Hittites, the Amorites, the Perizzites, the Hivites and the Jebusites" (Exodus: 3:8). The advantage of a model of God as a person is also that people can imagine a relationship to God in a way that is common and natural for them. They can pray to God, talk to God, love God and imagine being loved by God, etc. However, the model of God as a person should not be understood iconically but indexically (Neville 2013: 20–22). Indeed, God is not *really* a person. Linda Trinkaus Zagzebski suggested that we should not understand God as a person; rather it is *personhood* that is the quality of humankind that distinguishes them from other instances of nature and that the personhood of human beings is the most significant way in which humans are like God (2004: 203).

Hence, from the assumption that God created human beings as *imago Dei* and that human beings resemble God in their personhood, the assumption logically follows that God possesses personhood (which is not the same as to say that God is a person) (Zagzebski 2004: 209). A person but not necessarily personhood needs a body. Indeed, the Bible tells us that God is being, not a being, In Exodus (3: 14) God tells Moses that God is *ehyeh asher éhyeh,* which means "I am who I am or I will be who I will be". Hence as Ted Peters observes, the Old Testament uses the Hebrew verb 'be' (2013: 47. I will return to this discussion below because Zagzebski's suggestion has advantages for a way to understand God's action in the world.

When did God become a person, a being, ens? Understanding God as a person (or at least a being possessing personhood) can be dated back the New Testament. As I mentioned above, Jesus talked about God as our Father. Jesus' Father can be understood as a father, a person, but also as creator, the being (esse) that gave life to the world and everything in it. This will become clear below when God's actions in the world are discussed. However, God became a super-person during the middle Ages when Greek philosophy merged with Christian doctrine. Plato's world of ideas in particular became attractive for the Church fathers. This is understandable, knowing that Plato's world of ideas, which I would rather call the *ideal world,* is perfect compared to the imperfect world on earth and which he sees as a shadow of the real, proper world. Simply put, everything on earth is compared to gingerbread figures, baked in the same form but all slightly different, and hence not as perfect as the form itself, which is only found in the ideal world. The Church fathers adapted Plato's idea to represent God's creation. God created the world in accordance with God's own image. Hence, God was the perfect form and humans were represented by a tremendous variety of gingerbread figures, each one of them a little bit different from the others; hence, each one of them became unique. While I like this idea, the philosophical consequence could not be avoided; namely, an essence dualism between God and humans (and the whole of creation).

Hence, the question is: How, if God is so different from humans and the world, could God intervene in the world? How could God 'hear and answer' people's prayers? Hence the problem of theism is that God or ultimate reality is personified and put at an unreachable distance from the human world.

Another model of God or ultimate reality is modelling God as consciousness without objects, as is done in south Asian religious thought. The idea is that "where there is diversity in experience, especially change; there must be a deeper substratum of experience" (Neville 2013: 22). The advantage of such a view is that people can attain this consciousness (through meditation and other practices). Some argue that all of reality is consciousness. Hence, if everything is consciousness, then a connection between ultimate and everyday reality is possible.

Perhaps the best known example of a pure connection would be the 'Buddha mind'. This is a state of consciousness which is said to be pure and perfect. It is argued that when being in such a state there is no doubt of what is real and what is not. The idea is that people who reach this ultimate state of mind will no longer be attached to anything because they have become one with ultimate reality. Another advantage is that there is no dualism in the sense that there is no gingerbread and gingerbread form, since the reality as such is consciousness. However, there are also disadvantages with such a model. For example, life and death are seen as a stream of existence and transmigration. This implies that people should not pay too much attention to their earthly experiences, no matter whether these are good or bad, because they all belong to life's movement and belong to the impermanent nature of the world. This understanding of life led people to see themselves as subjects of change rather than subjects who can make changes. In turn, this led people to feel powerless and passive. While 'higher consciousness' may be obtained, there is no communication with God or ultimate reality; there is no hope for God's interaction in one's life and the world, and so forth.

Of course there are many variants of the model of ultimate reality as consciousness. My purpose is simply to present some models of God or ultimate reality as an introduction to the God-human relationship and God's actions in the world.

Another model of God or ultimate reality is the model of emergence, which is seen in East Asia. For example, the Dao themes of novelty and spontaneity and those of Confucianism have been modeled in the spirit of emergence. (Neville 2013: 23). Both Daoism and Confucianism rely on Dao, the source or the mother of the flow. The idea is that the flow emerges from a deeper reality that cannot be named. The principle is harmony between opposites (Yin and Yang). There seems to be a kind of double emergence. "[T]he emergence of temporal flow involving the temporal emergence through yin and yang, from something more basic than flow, beginning with the Ultimate of Non-Being, which has no qualities" (Neville 2013: 24).

This implies that every temporal emergence also includes non-temporal or eternal emergence. The ultimate reality is heaven to which people are related with respect to fulfilling their destiny, because destiny is what people receive from heaven. (Xinzhong 2000: 149). Ultimate reality is one but unites different aspects; "heaven with respect to its physical body, the Lord (ti) with respect to its being master, negative and positive spiritual forces with respect to its operation, spirit (shen) with respect to its wonderful functioning, and ch´ien with respect to its nature and feeling" (Xinzhong 2000: 142). Harmony between heaven and human is seen as the most important principle since "it denotes a moral correspondence between humans and their ultimate" (Xinzhong 2000: 174).

This harmony can be established through the Way (Dao), rituals/propriety (li), humaneness (ren) and virtue (de). Hence, education, virtue and self-cultivation become very important. While models of ultimate reality as consciousness focus almost entirely on the ultimate and hardly at all on material life, Confucianism focuses on both life here and now as well as on heaven and destiny. (Xinzhong 2000: 26). The advantage of the emergence model of ultimate reality is that ultimate reality is accessible by way of following one's destiny. Hence there is a relationship between humans and ultimate reality but it is a one-way relationship from humans towards ultimate reality. Similar to the consciousness model, ultimate reality does not interfere with human life or creation; different from the consciousness model, the model of emergence leads to something new. When harmony is obtained something new occurs, namely 'fulfilled destiny', whatever that is. Models of ultimate reality as consciousness, as well as models of ultimate reality as emergence, obviously avoid the problems that the personal model encounters, not least the problem of dualism, particularly encountered in theism; it avoids the problem of dualism all together. Since the aim is to present and discuss the possibility of a relationship between human beings and God and between God and human beings, I will now evaluate how fruitful the above models could be for such a relationship.

God's Action in the World

Whether and how God or ultimate reality acts in the world is the subject of much discussion within philosophy, religious studies and the discipline of science and religion. The question is not relevant if God or ultimate reality is understood as consciousness. Ultimate reality can only be experienced within one's consciousness. Neither is the question relevant when God or ultimate reality is understood as emergent processes; ultimate reality can only be experienced as one fulfills one's destiny. The question becomes relevant when God or ultimate reality is understood as personal, which is the case for the Abrahamic religions. Because I belong to the Christian cultural tradition, I will tackle the question from a Christian cultural perspective. I will therefore reframe the question into how God could act in the world from a Christian perspective. There is no simple answer and it also depends on how the divine is understood; within Christianity there are different ways of interpreting God. Take, for instance, deism (from lat. Deus). Those endorsing a deistic understanding of the divine believe that God created the world in all its aspects, but that once the world was in place, God withdrew from it. This implies that God does not intervene in the world; God does not act upon it. This view fits perfectly with the theory of evolution, because once the world had been created, the world and everything in it could only evolve by itself. God is still seen as a person or at least a being that possesses personhood, a creator. However, once creation was established, the world-human-God relationship ended. Hence the answer to the question whether God acts in the world is that God does not. The answer to the question how God acts in the world is irrelevant. There is no relationship between the divine and human beings (and the world) after creation has been established.

Another interesting understanding of the divine is known as pantheism. Pantheism means all things are divine. God is in the world. "The being of God and the being of the world are co-spatial and co-temporal" (Peters 2013: 52). Pantheists believe that everything comprises an immanent God. Hence, they do not believe in a distinct personal God. Nothing exists outside of God. God *is literally present* in everything. The answer to the question whether God acts in the world is yes. The answer to the question of how God acts in the world is that God works in the world from within, from within every possible detail of the universe. This means that the divine can be encountered directly in everything we experience; God can be seen in everything, in our natural, social, and interpersonal environment and within ourselves.

Let us now turn to the view that is embraced by most Christians but which is also the most philosophically problematic in relation to divine action in the world, namely classical theism. The classical theist's understanding of the divine is that God is entirely free and separate from the world. God is simple; in other words, God is one. God is beyond composition; God has no parts, be it material or metaphysical parts. God is incorporeal, which can be interpreted in different ways. Firstly, incorporeal can be understood as God is disembodied, without body, secondly that God is entirely spiritual, and thirdly that God is ethereal.

The first understanding implies that God cannot be a person because persons typically have bodies. The second and the last understanding may imply that God could be a person, namely a spiritual or non-earthy person. The classic theistic God is immutable or changeless, impassible, entirely external (spaceless and timeless); entirely transcendent and is without existence or essence. God is omniscient, omnipotent and caritas. Through God's caritas creation is sustained. The answer to the question whether God acts in the world is obviously yes. The answer to the question of how God acts in the world is, to say the least, problematic. Simply put, if God is entirely other and entirely separated from creation, how can God act in creation? How could a spaceless and timeless God act in a space-time dependent world?

One answer is by way of miracles (which feature in both the Old and New Testaments). Classically a miracle is understood as God suspending or interrupting the laws of nature in order to bring about a certain course of events. Arguments against the existence of miracles are that natural laws cannot be violated otherwise they would not be natural laws; and that miracles are unfair because they only happen to some and not others. The latter argument implies that if God intervenes by way of miracles God would be amoral and this does not fit well with the understanding of God as caritas. Maurice Wiles (1923–2005) wonders why there were no miracles during the great human tragedies:

> If the direct action of God, independent of secondary causation, is an intelligible concept, then it would appear to have been sparingly and strangely used. Miracles must be by definition, relatively infrequent or else the whole idea of the laws of nature [...] would be undermined, and ordered life as we know it would be an impossibility. Yet even so it would seem strange that no miraculous intervention prevented Auschwitz or Hiroshima, while the purposes apparently forwarded by some of the miracles acclaimed in traditional Christian faith seem trivial by comparison (2012: 66).

Christine Overall argues similarly, "In choosing to favor just a few individuals, God shows himself to be arbitrary in his beneficence to some and cruel and unfair in his neglect of others" (2003: 131). Besides the ethic concern, the theistic understanding of God suffers from the problem of theodicy which defends God's goodness and justice in the face of the existence of evil. In brief the problem is as follows:

1. Is God willing to prevent evil, but not able? Then God is not omnipotent.
2. Is God able, but not willing? Then God is malevolent.
3. Is God both able and willing? Then why is there evil in the world?
4. Is God neither able nor willing? Then why call God God?

However, can the occurrence of miracles be entirely dismissed? Thomas Jay Oord does not think so. First of all, people of different religions have witnessed miracles. More importantly, he argues that in order to know that a natural law has been violated, one needs to understand the laws of nature profoundly, which "no one seems to do" (2015: 194). Furthermore, nature has its irregularities. Not entirely wanting to refrain from theism, some theologians and philosophers have tried to solve the problems by way of changing some aspect of the theistic understanding of God. Some suggest that God cannot be understood as a person because this implies having a body and belonging to the world; God is beyond the world and hence cannot have a body and hence cannot be a being. They argue that God is beyond or without being (Stenmark 2015: 206). Stenmark has called such an understanding of God 'Alterity theism' (2014: 206).

> God is therefore not only not a 'being,' he is also not subordinate to, nor contained by, the Being of beings. The whole panoply of metaphysical procedures – rational argument, empirical evidence, experiential inference and so forth – simply cannot apply to God. If they are, they are treating God as an 'object' or 'thing' within the realm of Being, and the conclusion of such an investigation will inevitably be that such an 'idol' does not exist (Hyman 2010: 175 in Stenmark 2015: 206).

What I found interesting is that God should not be understood as 'a being' (ens) but as 'being' (esse). As mentioned above, after all, God's name is Yahweh, which refers to 'being', rather than to 'existing'. Yahweh means 'I am what I am, or will be what I will be' and not 'I am a person'. However, that something does not exist in the sense of being measurable or scientifically verified (mathematically or empirically) is not to say that it cannot be real (as I pointed out above).

Also mentioned above, Zagzebski argues that God cannot be understood as a person. Her solution is to understand God in terms of personhood. A person but not necessarily personhood needs a body. Personhood is hence seen as esse rather than ens. Charlene Burns suggests that divine personhood is the relationship "in which God manifests a deep feeling for creation" (2002: 14). God loves creation and shows mercy, forgiveness, care, compassion, empathy and so forth.

Theism, by way of changing the concept of person into personhood, solves some of the problems of classical theism in relation to divine action in the world. In my opinion apprehending God in terms of personhood is fruitful for understanding divine empathy as well as the doctrine of the *imago Dei*. If we accept the view that God possesses personhood of which empathy is one quality; that divine empathy is directed towards all of creation; and that human beings are created in God's image, which is mirrored in human personhood, it becomes plausible to argue that divine empathy can be mirrored in human personhood at least at a low; and furthermore, that divine empathy at a low level is innate in non-humans as well as in humans not only as a capacity for survival of the species but also as a precondition for developing high level empathy such as selfless altruism and agape. Perhaps empathy is not only the core subject matter of mindreading but also the hallmark of the God-world-human relationship. Perhaps one could argue that the longer and more intensely people have lived with their belief in God, the more their personality comes to mirror divine qualities such as empathy. What the understanding of God in terms of personhood does not solve is the problem of the God-creation dualism. This is so because divine personhood is still put over and above human personhood.

Another alternative for solving the problems encountered by theism is to endorse a panentheistic view of God or ultimate reality. Roughly, those defending this view argue that the entire creation exists within God without exhausting God's being. As Ted Peters says, "there is a little bit of God left over, so to speak" (Peters 2013: 57). Another way of putting it is to say that God is in and over creation. God is in creation and also embraces it. Some panentheists compare the God-world relationship to the mind-body relationship, meaning that the world is God's body and God is the mind of the world. However, this analogy needs further exploration, because the mind-body relationship is understood in different ways; i.e. as two different realities (dualism), as one and the same reality (monism), as two sides of the same reality (complementarism), or as a threefold reality (as I suggested above). Hence, whether and how God acts in the world will depend on how the mind-body relationship is understood. Understood in a dualistic way it encounters the same problems as classical theism. Understood in a monistic way, God and creation would be one and the same reality, which is pantheism rather than panentheism. The same counts for the complementary understanding of the relationship. The most promising understanding is the last one, because there is one aspect of that reality that transcends the other two. I will return to this below.

What also needs to be said concerning panentheism is that panentheists do not accept the notion of *creatio ex nihilo*. Panentheists believe in a continuing creation, *creatio continua*. Because of the God-world interdependency, the price panentheism pays is that God's power is restricted. We need to answer the questions: Does God act in the world? Yes. How does God act in the world? The answer is from within and beyond the world. How, more exactly, God acts in the world is open to different interpretations. I shall investigate this problem further by first analyzing the problem of divine action in the world in terms of temporality. If God is understood to be atemporal, how could God act in temporal world? Secondly I will look at some models of how God could interact with creation from an interdisciplinary point of view.

Atemporal Divine Actions in a Temporal World

The question raised is: If God is understood to be timeless or atemporal how could God act in a time-dependent or temporal world? To answer this question we need to go back in history to understand how this problem arose. I will focus on three possible explanations provided by Anne Conway, Paul Helm and Nelson Pike. But first, what is meant by the term atemporal?

Atemporal duration, it is argued, is existence possessed completely, all at once, present entirely to its possessor. In contrast, temporal duration is existence possessed with radical incompleteness, a bit at a time, mostly already lost to its possessor or not yet possessed (Stump and Kretzmann 1987: 219).

That God has been apprehended as being atemporal has its origin in the Old Testament. In the Old Testament God acts, reacts, has a change of mind, and is always present with the people. However, when reading Psalm 90, for example, the impression given is that God is timeless.

> Lord, you have been our dwelling place throughout all generations.
> Before the mountains were born or you brought forth the whole world, from everlasting to everlasting you are God (Psalm 90).

Still it is not this notion of eternity that has caused the philosophical problem of how a timeless God could act in a time-dependent world. According to Anne Conway (1631–1679), the fact that the will of God is eternal implies that creation is necessarily immediate, without any time interval between God's will and creation; creation is the result of the will to create. She wrote "[a] nd yet it cannot be said that creatures considered in themselves are coeternal with God because then eternity and time would be confused with each other" (1996: 12). In Conway's eyes, creation, as well as time, have their beginning in God's eternal will.[4] Paul Helm argues, as many others have done, that it was Saint Augustine who created the problem by putting God beyond time and creation. He writes, "The doctrine of God's timelessness seems to have entered Christian theology from Neo-Platonism, and there from Augustine to Aquino it reigned" (1988: 1). If, he argues, we keep to the biblical understanding that God is *in* time, we will not have this problem, because the biblical understanding is that God acts, reacts and changes. Nelson Pike argues similarly when he writes that the doctrine of an atemporal God was introduced because, at that time Platonism was fashionable and its theories reflected a systematic elegance. (Pike 1970).

Eternity not only meant *always* or *in all times,* but was now understood as *timeless.* Temporality belonged to the universe and everything in it. God, beyond the universe, on the other hand, became perfect or absolute; and therefore God could not be *in* time. God became timeless or atemporal. Saint Augustine put God far above the creation, implying that the physical and the heavenly could never meet. Once introduced within theology, the doctrine of an atemporal God gathered momentum. God is understood as the creator and infinite spirit. Furthermore, God

[4] Anne Conway refuted (amongst other theories) the dualism of Henry More and Descartes.

created *ex nihilo*. If God were finite instead of infinite, God would be dependent on someone or something else to create, which implies that God could not be the creator. If creation had not occurred *ex nihilo*, it would have occurred through stuff which already existed, and since this stuff would have been infinite, God could not have created everything as it is believed God did. Timelessness became an attribute amongst others that guarantees the attribute of consistency; something which is necessary in order to explain the difference between the creator and the created. Paul Helm quotes Anselm of Canterbury (1033–1109)[5] when he writes:

1. Whatever is created is finite.
2. Whatever is finite is mutable.
3. Whatever is mutable is in time. Therefore,
4. Whatever is created is in time.
5. Whatever is the creator is infinite.
6. Whatever is infinite is immutable.
7. Whatever is immutable is outside time. Therefore,
8. Whatever is the creator is outside time. (1988: 18)

We have a timeless or an atemporal God on the one hand and a time-dependent or temporal universe on the other.

This is so, as long as we understand time in a Newtonian manner, because Newton's worldview is dualistic and his temporal order is described as having a past, a present and a future. Time itself is absolute and moves constantly without being affected by external features. However, the problem is still not solved either if we apply the Einsteinian understanding of time, where Einsteinian time describes a four-dimensional worldview. God, apprehended as atemporal, would still be beyond the Einstein space-time world because we are still confronted with a dualistic view of reality; i.e. due to classical theism, God and the world are hopelessly separate.

Nevertheless, Eleonora Stump and Norman Kretzmann think otherwise. They argue that understanding God as an atemporal actor in a temporal world needs to imply a time paradox. The problem, according to them, lies in a misunderstanding of the concept of eternity. One misunderstanding is that eternity is equated to endless elongation in time. Another misunderstanding is that eternity is identified as atemporality, which means an isolated static moment. Taken from Anicius, M.S. Boethius (480–524), "Aeternitas igitur est interminabilis vitae tota simul et perfecta possessio" (Boethius 1999; Stump and Kretzmann 1981: 431).[6]

Stump and Kretzmann put their finger on four elements in Boethius' definition of eternity causing incoherence.

1. What is eternal has life.
2. The life of an eternal being is infinite (there is no beginning and no end).
3. This implies an infinite time extension.

[5] See further Anselm of Canterbury 2008.

[6] Eternity and absolute ownership by an infinite life.

4. This implies that to be a being having both absolute and infinite life is to be atemporal. A being that owns his/her life all at once cannot be temporal but needs to be atemporal.

According to Boethius, eternity is a form of existence that cannot be reduced to or is incompatible with time. Stump and Kretzmann argue that the terms atemporality and temporality are apprehended from within the wrong frames of reference. This is an important remark. For example, atemporality is explained in terms of an infinite present and is hence understood as temporal, as a *moment*. For example, let us try to divide an infinite extent of time into parts. The problem is that the extent consists of lengths, which consist of lengths, which consist of lengths, in a never ending manner. The Greek philosopher Zeno (490 BC–430 BC)[7] has already pointed to this particular problem; i.e. the problem of dividing an infinite distance into a finite one. In order to reach a certain distance one has to cover half of the distance, but in order to cover half of that distance one has to cover half of the half of that distance, and so on, to infinity.

The problem is that within finite time one cannot cover an endless distance; hence, the result is that one will never reach one's goal.

> If a runner is to reach the end of the track, he must first complete an infinite number of different journeys: getting to the midpoint, then to the point midway between the midpoint and the end, then to the point midway between this one and the end, and so on. But it is logically impossible for someone to complete an infinite series of journeys. Therefore the runner cannot reach the end of the track (Audi 1999: 987).

Stump and Kretzmann therefore point to the illusion of the paradox, and maintain that we can only divide infinite distances (or time) at an *abstract* level, not at a *physical* level. We can only explain the concepts of infinity and finite time from their specific frames of reference. To repeat the incoherence, an infinite being's duration cannot be divided into sequences. From the point of view of eternity, there is no earlier, no later; nothing belongs to the past, the present, or the future.

The second incoherence concerns atemporality and life. Life is understood to incorporate a succession of time. Life is associated with processes and processes are time-dependent. The consequence then becomes that an atemporal being could not possess life because atemporality is incoherent with life. Stump and Kretzmann tackle this problem with the help of Thomas Aquinas' (1225–1274) idea that God is divine consciousness in the sense that God contains an infinitely multiple consciousness and that God as a whole is always aware of all His consciousness (Aquinas 1991, in Stump 2003). They then distinguish between conscious acts that imply time and those that do not. For instance, to read and think imply time, but to know does not. To wish implies time but to will does not. Similarly, to perceive implies time but to be conscious does not. The point they want to make is that all types of consciousness possess life; even types of consciousness that do not possess all the attributes living beings normally have (Stump and Kretzmann 1987). At first sight it

[7] Zeno: see Audi 1999: 987–988.

seems that they have solved the problem, and to a certain extent they have. However, and unfortunately, their solution implies dualism.

Indeed, it implies a dualism between acting and being (esse) or at least between physical actions and conscious actions.

To be sure, the assumption with which Descartes begins has come to seem very questionable; his own conception of (conscious) minds as independent substances, fully unified within themselves, and wholly distinct one from another, fails to do justice to the empirical facts. And even in the absence of those facts, it is surely a shortcoming of his view that fails to accommodate the logical possibility of simultaneous, overlapping phenomenal perspectives (Lockwood 1989: 94).

What catches a philosophical mind is the sentence '*… fails to accommodate the logical possibility of simultaneous, overlapping phenomenal perspectives*'. The atemporal being is conscious being (or consciousness in itself, Consciousness with a capital C) that only acts mentally; while the temporal being acts both physically and mentally. The question is raised again, how can an atemporal Consciousness act in a temporal conscious-physical world? Stump and Kretzmann try to solve this problem by introducing the concept of co-existence. However, does co-existence not imply simultaneity? They make a division between temporal co-existence and atemporal co-existence. Because atemporal existence cannot include any time-order, they define atemporal existence as simultaneous existence or atemporal actions within the same infinite presence.

Suppose we designate the ordinary understanding of temporal simultaneity T-simultaneity: (T) T-simultaneity is existence or occurrence at one and the same time.Then we can proceed and construct a second species of simultaneity, a relationship obtaining between two eternal entities or events: (E) E-simultaneity is existence or occurrence at one and the same eternal present. (Stump and Kretzmann 1981: 435).

The problem is that in order to show that it is plausible for a timeless God to act in a time-dependent world, Stump and Kretzmann need to define simultaneity between two relata, one of which is temporal and the other atemporal. Hence, they establish such a relationship by highlighting what T and E have in common, and this turns out to be simultaneous existence or events. However, in order to establish TE, simultaneous existence or events need to be supplemented with some type of extension coherent with both T and E. Since TE includes an atemporal relatum, a TE relationship needs to refer to one and the same present, rather than to one and the same moment. Indeed, events may happen simultaneously but do not need to meet, they can happen in parallel. Furthermore, TEs consist of two independent entities (a temporal and an atemporal one) of which neither is reducible to the other. Inspired by Einstein's Special Theory of Relativity (STR) Stump argues, "[e]vents occurring at different places which are simultaneous in one frame of reference will not be simultaneous in another frame of reference which is moving with respect to the first. This is known as the relativity simultaneity" (Stump 2003: 140).

Hence we are forced to define simultaneity as existence or occurrence at the same time within the frame of reference of a given conscious observer. Since God is understood as consciousness, the problem of divine atemporal action in the temporal world seems to be resolved. Is this the case? Let us take a closer look.

1. If x and z are temporal entities, they co-exist, if and only if, there is a time-zone in which both x and z exist.
2. If something exists atemporally, its existence, regardless of its infinite extent, is wholly present.
3. It follows from this that the entire life of every atemporal entity co-exists with every temporal entity in every moment of its existence.
4. Hence, seen from a temporal frame of reference, the atemporal is always present in all its potential.
5. Seen from an atemporal frame of reference, all times are present and co-exist with the entire atemporal existence. (Stump and Kretzmann 1981: 441) (Fig. 8.1).

Illustration of TEs according to Stump and Kretzmann. Two infinite parallel horizontal lines, the upper one representing eternity, is entirely and uniformly a strip of light (where light represents the present), while the lower one, representing time, is dark everywhere except for a dot of light moving steadily along it. [...] the light, in the representations of eternity and of time, should be interpreted as an indivisible present. For any instant of time as that instant is present, the whole of eternity is present at once; the infinitely enduring, indivisible eternal present is simultaneous with each temporal instant as it is the present instant (Stump and Kretzmann 1987: 219).

Thus far, Stump and Kretzmann were able to provide an alternative understanding of how God, being atemporal, could act in a temporal world. The idea is that God hears prayers, which implies divine action. Being aware that such action implies a time-interval, they argue that they solved the problem by establishing a TE simultaneity.

> [...] consider the case of Hannah's praying on a certain day to have a child and her conceiving several days afterward. Both the day of her prayer and the day of her conceiving are ET-simultaneous with the life of an eternal entity. If such an entity atemporally wills that Hannah conceive on a certain day after the day of her prayer, then such an entity's bringing it about that Hannah conceives on that day is clearly a response to her prayer, even though the willing is ET-simultaneous with the prayer rather than later (Stump and Kretzmann 1981: 450–451).

However, from what has been pointed out earlier in relation to Stump and Kretzmann's use of consciousness, as well as their division of conscious activities,

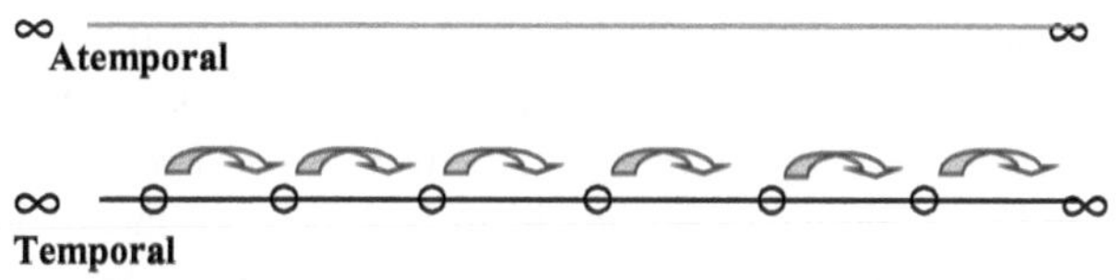

Fig. 8.1 Stump's and Kretzmann's reflections on time

which risks dualism, it can only be argued that God acts alongside and not 'in' the world. It is not enough to show mutual action. Co-existing does not necessarily imply co-working. For instance, at this very moment, my husband and I co-exist in regard to place (our house) and time (3 p.m.). However, we do not co-work. He is in his office working with his stuff, and I am in mine writing this book. Furthermore, they depart from classical theism by arguing that God should be understood as consciousness (be it Consciousness that intervenes in the world by way of divine Will) rather than as a person. I believe this to be a good move, but it does not save classical theism. In my opinion they should have taken a step further; namely, to move from Einsteinian time to quantum time. Can the problem be solved?

Let us return to the problem of how can God be understood as being timeless and also act in a time-dependent world? We have seen that within quantum physics it is the observer who plays the leading role. It is the observer who has access to a diminished (collapsed) reality. As mentioned earlier, we do not have access to the whole of reality (the superposition). However, we can imagine that God has, if we accept the understanding that God is omniscient and timeless. In the language of quantum physics, God would have access to the superposition itself, including all its possible states. For example, God could have access to both the great fire in Rome, in the year 64 AD, and to you reading my book in 2016. Suppose the impossible, that we could reconstruct the superposition God has access to, we would see a state (*eigenvalue*) in which Rome burns and in which Rome does not burn; in which you read this book and in which you do not care about reading this book; as well as all other possible states. There is no time. One might perhaps say that God who is timeless encompasses the whole superposition. If we reconstruct what you have access to, and suppose you actually are reading this book, then you have collapsed the superposition into 'reading this book just now' and you will have put a time label on 'reading just now'. You will not see a burning Rome, but of course you will see other states related to you reading the book. Suppose now that you discover the book *The Great Fire of Rome* on the shelf. What happens is that you collapse the superposition again from within your frame of reference, putting a time label on that observation. Suppose now that you take the book and immerse yourself into the historical event of the great fire of Rome; you will put the time label now on it (i.e. your time label at the time of reading this book). This means that God, understood in a quantum mechanical sense (rather than in a Newtonian or Einsteinian sense), can be timeless and act in the world; but the action is through you. To me, this line of thinking fits well with the Christian understanding that it is we who need to will God. God does not force anything upon us.

Let us look at the following example. Suppose Jacob has a cold at time t. Suppose God cures him at time t′. Obviously there is a time interval between t and t′. Hence, understood in a classical way, God cannot cure Jacob, because that would imply that God has to be in time in order to act upon Jacob. Jacob's personal biography can be presented as Ψ (…). (There is nothing between the brackets because there is nothing to be remembered yet). When Jacob lies ill in bed, he correlates with the state of illness but also with the state of wellness (he remembers both being sick and well). In a manner of speaking, he is part of the superposition together with all possible states of biography. Suppose Jacob observes, at time t + 1, that he feels sick; call it Ψ+ 1 (…,Ψ). Following Lockwood then, this *eigenstate* will be followed by a phenomenal perspective, Ψ+2 (…,Ψ, Ψ + 1), that is associated with a *memory* of this impression of feeling sick. What happened between time t and time t + 1 relative to his biography (which includes consciously observing feeling sick) is a transition in the relative state of having been healthy to becoming sick. In the state Ψ (…) Jacob is both sick and well. Suppose further that at time t + 1 he observes that he feels well again; call it Ψ' + 1 (…,Ψ).[8] Following Lockwood again, this *eigenstate* will be followed by a phenomenal perspective, Ψ' + 2 (…,Ψ, Ψ'+1) that is associated with a *memory* of this impression of being well. Then, relative to his biography, the transition from t to t + 1 is associated with a transition in the relative state of having been sick to becoming well. Jacob now feels healthy Ψ+ 2 (…, Ψ).

Hence, within a quantum physical perspective, God does not need to be in time; it is enough that Jacob is. It is enough that Jacob observes that he is cured at time t′; that he has observed the alteration from having been ill to having become well. From a God's-eye perspective (the perspective of the superposition), Jacob is in a state of illness, in a state of wellness, and all other possible states. There is no time. If Jacob, say after praying or meditating, observes that he has been cured, it is Jacob who collapsed the superposition into wellness at time t′. In my opinion, a quantum physical understanding provides an alternative view as to how God can act in a temporal world while being atemporal. However, this is not to say that there are no other problems with a quantum physical worldview.

One problem is of course that God could not intervene in the world if not asked to intervene. Hence, such an understanding of God's action in the world is not deistic, pantheistic or theistic. However, might this understanding be relevant for panentheism?

Divine Action in the World

While the previous section tackled the problem of divine and actual time, this section will look at the problem of divine action, not specifically in relation to time. I will present and discuss some alternatives. One suggestion is provided by Robert Cummings Neville who established the idea of God or ultimate reality as "the ontological creative act" (2013: 24–29). Another suggestion comes from Arthur Peacocke who established a panentheistic model of the relationship between God and the world based on how the world is comprized. While looking at panentheism as an option, I will also briefly present Nicolaus Cusanus' pre-panentheistic view. Finally, leaning on my idea of the three-fold or threefold self, and inspired by Peacocke, Cusanus, Philip Clayton, Zagzebski and Celia Deane-Drummond, I will suggest my own panentheistic model.

[8] The prime ′ indicates that this is a different eigenstate of the relevant set of compatible brain observables.

The Ontological Creative Act

Neville's idea behind the ontological creative act is that he wants to depart from "the determinacy of personhood, consciousness and emergence to something that is beyond determinacy [...]" (Neville 2013: 24).

To be determinate is always in relation to something else. Something can be determinate in relation to x and y, but perhaps not in relation to p and q. To be determinate is also to be conditioned, e.g. causally conditioned. Furthermore, to be determinate implies having essential features. Neville argues, "[a] determinate thing is a harmony of essential and conditional features" (2013: 25). By 'harmony' he means that all the features simply fit together. Obviously, because to be determinate is always in relation to something else, there have to be at least two determinate harmonies. Neville talks about a plurality of determinate harmonies. The harmonies then possess "four transcendental traits of determinateness described as cosmological ultimates" (2013: 25); or, as I would say, communicative traits, as a result of the fitting together; components that are formed in the harmony's pattern also need to be harmonies, and are established in their existential location in relation to other harmonies; lastly, there is the capacity to gather the components and the form in their existential location. That different determinate harmonies can be related to each other supposes that they "exist within an ontological context of mutual relevance [which] must in itself be indeterminate" (Neville 2013: 26). To Neville, the indeterminate ontological context of mutual relevance is an ontological act of creation. "The act is indeterminate except in giving itself the nature of being creator of the world of determinate things" (2013: 26). The ontological act of creation or ontological ultimate reality (the creator) also has three specific features: Firstly, the act itself or the making; secondly, the world as the terminus of the act or the made; and thirdly, the nothingness or *creatio ex nihilo*. Importantly, Neville defines ultimacy or the ultimate as the "final condition beyond which there is nothing more" (2013: 28). Hence, he continues, the four transcendental traits are ultimate realities. For example, having form is an ultimate reality, in the sense that there is no other form beyond it; what is beyond is formless chaos. In other words, these conditions and features are in the boundary conditions and features of the world.

The question now is how we could understand a relationship between the ontological creative act (the creator) and the world. In other words, how could God act in the world?

Neville does not directly tell us whether God is active and intervenes in the world. He only says that "sometimes people have signs that express the finite/infinite contrast [and calls this] religious engagements of the ultimate" (2013: 28). The ultimate conditions are understood to be both finite and infinite (e.g. form and chaos). There is harmony between finite and infinite contrasts (the finite and the infinite fit together). It is these harmonies that are apprehended as signs of divine action in the world. Perhaps only looking at possible divine action, we should also look at action within the world itself and investigate whether there could be an opening for divine action from within the world. There are some proposals that I find attractive and which inspired me to further develop these models of divine action in the world.

Panentheism

Indeed, there are different ways to look at the universe. One way I found attractive for my purpose is the way in which Arthur Peacocke understands the universe. To him, there is only one universe, something science, especially mathematics in elucidating the entities, structures, and processes of the world, continues to reinforce (2004: 138). All entities emerge from that world. We live in a complex world consisting of complex hierarchies. "Each level in the hierarchy of complexity is specific to what is distinctive about it. [...] [s]ociological, psychological, and biological concepts are characteristic of their own levels and quite different from those of physics and chemistry" (Clayton and Peacocke 2004: 138). These different levels have a different degree of complexity. The physical level is the least complex one, followed by the level of living organisms. More complexity is found in the behavioral level of the living organisms. The level possessing most complexity includes culture, philosophy, theology and religion. The hierarchy Peacocke refers to should not be understood in terms of higher-level states of reality being reducible to lower-level states of reality, a doctrine to which reductionist philosophers of science subscribe.

For example, "[...] the laws of such 'higher-level' sciences as psychology and sociology are reducible to the laws of lower-level sciences – biology, chemistry – ultimately to the laws of elementary particle physics" (Putnam 1994: 428). In other words, these philosophers maintain that each science, physics, chemistry, biology, psychology, sociology, is supposed to be reduced to the one below and ultimately to physics. What Peacocke argues is that there is only one universe, in which, through the mutual affection of the hierarchies and the parts, new and distinct kinds of realities emerge. It is important to note that Peacocke's world includes the natural as well as the ultimate world. God, he maintains, is the ultimate ground of all necessities and possibilities, which are two necessary parameters for a self-organizing world. Peacocke's world is hence a closed system in which God also belongs. He writes,

> God could affect holistically the state of the world-system. Thence, mediated by the whole-part influences of the world-system on its constituents, God could cause particular patterns of events to occur which would express God's intentions. These latter would not otherwise have happened had God not so specifically intended (2002: 51–56).

Hence, God's interaction with the world-system is initially on the whole-level. God's interactions then may be trickled down through the whole-part-influence action, through the hierarchies of complexity to the level God intended to affect. God's interaction with the-world-as-a-whole and the whole-part influences can be termed as top-down causation. The parts-whole effect means that the parts, as it were, transfer both initial and altered effects to the higher hierarchies of complexity which, in turn, might be transferred to God.

The question raised by Philip Clayton is: Does God in fact engage in any such focally intended actions? Does God's ongoing creative involvement in fact make any difference to the way anything happens, compared to the way things would happen if God were not involved and the universe were simply allowed to run by itself? (Clayton and Peacocke 2004: 263) Peacocke answers that God (continuously) provides some input to 'the whole', which through "a long chain of trickle-down causes [makes] particular things happen at particular places" (Clayton and Peacocke 2004: 263).

Clayton is not entirely satisfied with Peacocke's panentheistic reasoning. He argues that there has to be an element that is pure spirit insofar as it participates in the divine essence. God did not create an open-ended universe and then merely hope for a final consummation that would be to God's liking. God must also "draw the world unto himself" through some sort of causal activity. This activity he understands as God's providence. This allows for actual divine influence, instead of the domino-effect idea of Peacocke (Clayton and Peacocke 2004: 264). Clayton's model of panentheism is emergent panentheism. The emergent concept is a viable means for expressing the relation 'in' or 'is internal to' (metaphor). It forms a logical relationship, namely, belongs to, is a member of, etc. This means that the relationship is a more logical than spatial one. The nature of the 'in' or self-inclusion relation changes as one moves up the hierarchy of complexity in nature" (Clayton and Peacocke 2004: 88–89).

Peacocke does not defend pantheism; rather he argues for panentheism, saying that God is in the natural world but is not equal to that world (which would be pantheism). God depends on the natural world as well as the natural world depends on God, but God cannot be reduced to the natural world. God depends on the natural world in the sense that God could not have not created it but this does not mean that God could not have created a completely different world. God's being is distinct from creation in "a way that we are not distinct from our bodies" (Peacocke 2002: 109).

This is, however, not the same as some panentheists argue; that the world is God's body and God is the mind of the world.[9] The *en* in Pan-*en*-theism stands for a relationship between the world, human beings and God. Celia Deane-Drummond asks what the purpose of a relationship between God, the world and human beings might be. Her answer is that the purpose is *friendship*. "Jesus as the Wisdom of God", she argues, suggests that the relationship between God, the world and human beings should be understood in terms of friendship, which is also a feature of personhood. God, being the friend of the world and humans, participates in the distress

[9]As mentioned above, such a view is problematic, because the mind-body relationship is understood in different ways, i.e. as two different realities (dualism), as one and the same reality (monism), as two sides of the same reality (complementarism), or as a threefold reality (as I suggested above). Hence, whether and how God acts in the world will depend on how the mind-body relationship is understood.

of the world and of human beings but also in the world's development, and in human accomplishments and creativity (Deane-Drummond 2004: 259).

There are different understandings of panentheism: for instance, Niels Henrik Gregersen defines three varieties, soteriological, expressivist and dipolar panentheism. Soteriological panentheism dictates that the world's being 'in God' is not taken as a given, but as a gift. It is only by the redeeming grace of God that the world can dwell in God. This implies that not everything can be or is in God; for example, wickedness and sin are not. Soteriological panentheists also hold that self-revelation of divine love is not found everywhere in the world; only here and there. The 'all' of pan-en theism is qualified in terms of love and beauty. Truth exists in God, but not evil, because of the idea that not everything can be in God. It seems that soteriological panentheism accepts the idea that the world, but not all of it, dwells in God. It does not regard the possibility that God also dwells in the world. In other words, it is different from the panentheist idea that the world is in God as well as God is in the world.

Expressivist panentheism, which goes back to early-nineteenth century German idealism, claims that the divine Spirit expresses itself in the world by going out of God and returning to God, enriched by the experiences of world history. Only when the history of the world has been completed and sublimated in God will the circle of divine self-expression and self-return be closed. This is the view that God or ultimate reality is in everything that is, both in the good and the bad. While those models are interesting, the model of interest here, and which is most in line with Peacocke's panentheism, is, in my opinion, a combination of the expressivist and the dipolar panentheistic models.

In the dipolar model God is understood as being timeless, beyond space and self-identical in some aspects, while in other aspects God is understood as temporal, spatial and affected by the world (Gregersen 2004: 20–21). To Peacocke, God is related to the world as 'more than' a personal agent; God is also creator, giving the world its existence and infinitely transcending it. Peacocke uses terms like supra-personal or transpersonal to denote God's personality. Furthermore, God knows all that is logically possible to know about the world; God knows all the patterns of the world as well as all the events in it. (Clayton and Peacocke 2004: 150–151). According to Peacocke, God acts on the-world-as-a-whole but there is also a whole-part influence and parts-whole effect. (Peacocke 2002: 51–56). Peacocke uses the term whole-part influence to describe all the ways in which a system-as-a-whole, operating from its higher level, causes effects to its constituent parts, the lower levels. What the parts are and which functions or activities they have is due to their incorporation into a system-as-a-whole; they would not be or behave as such were they not. This is what it is to be a self-organizing or dissipative system. Thus far we have a top-down-bottom-up causation between God as a transpersonal creator and the natural world. This view is not entirely new.

Nicolaus Cusanus, a fifteenth century German cardinal, philosopher and scientist, may not have called himself a panentheist. Nevertheless, contemporary theologians have seen him as the anticipator of panentheism. Cusanus' view has its roots in his longing to harmonize antagonisms, *Coincidentia oppositorum* (Sandbeck

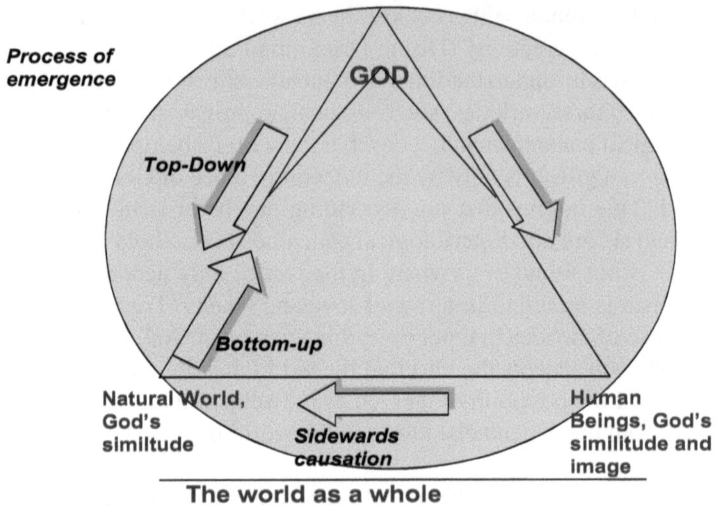

Fig. 8.2 Panentheistic model

2007: 76). God, the world and human beings are not separate authorities. There is merely a structural tension between these authorities (Sandbeck 2007: 77). The world is seen as the creator becoming visible, as God's similitude. The world is God presented as a multitude of 'real things' but the real things (as a whole or in part) are representations of God. So far, Peacocke's panentheism is in line with Cusanus'. The difference is that even though human beings are similar to the natural world, God's similitudes, they are also God's images, *Deus secundus*. Human beings create rational things. The world now becomes the place where God and humans meet: God shows Godself to humans through real things and humans are able to recognize God by way of rational things. The world is the mediating authority. Reality is a dynamic process: God, the world and human beings are continuously changing due to this *threefold process* (Fig. 8.2).

Combining Peacocke's and Cusanus' panentheism, God still has all the characteristics of a theistic God but, since God is also in the world-as-a-whole, some powers of God are reduced. This means that the world-as-a-whole includes struggles, challenges and even evil by necessity, but these nevertheless sustain the natural order process of emergence. There is *top-down-bottom-up causation*, by way of a complex whole-part and complex part-whole influence. By way of rational acts, human beings transform themselves into the natural world without reducing themselves to it since humans are both God's similitude, similar to nature but also God's image, which nature is not. It is now time to present my own panentheistic view which I call *three-fold or threefold panentheism.*[10] Since there are a variety of types of panentheism, the question is which 'type' would be the most fruitful? At least the following elements are needed for a panentheistic model.

[10] My idea was initially suggested in Runehov 2009b.

1. There is only one emergent universe (EU) comprising both ultimate reality (UR) and natural reality (NR).
2. God created the world from within Godself and interacts with it from within the world through the nested hierarchy of the world by way of multiple causations.
3. God is immanent and transcendent in relation to the world and human beings or, as I prefer to describe them, emergent selves (ES). $UR > (NR \cap ES)$.
4. The threefold panentheistic view includes the doctrine of the *imago Dei*; humans are both God's similitudes and *imago Dei*.
5. From this it follows that the world becomes the mediating authority between God and human beings.

The first, second, and third elements are covered, among other models, by Arthur Peacocke's panentheistic model, which implies that there is only one closed universe that includes both the natural and an ultimate world; that God is immanent and transcendent in relation to the world and human beings. Clayton's panentheistic models especially also emphasize that there is a process of emergence. God is relationally in the world but is not the sum of the world and, simultaneously, human beings are relationally in God. I wish to establish that the emergent universe is related in the following way. The emergent universe (EU) comprises ultimate reality (UR) and natural reality (NR) between which there is mutual causation; nevertheless, ultimate reality is bigger than natural reality. Hence, we can write,

$$EU(UR \leftrightarrow NR); UR > NR.$$

The justification for including elements 4 and 5 in the panentheistic model is as follows: Nicolaus Cusanus' panentheistic model comprises the concept of the *imago Dei* (Watts 1983). As Wentzel van Huyssteen argues: "For Christian theology one of the most crucial questions today should be whether there is a way in which we may rediscover the canonical function and orienting power of a concept like the *imago Dei* without retreating into metaphysical abstractions [and that would] facilitate[e] interdisciplinary reflection" (Van Huyssteen 2006: 116). The doctrine of the *imago Dei* is perhaps best be understood in the following quotation: "Humans mirror God's divinity in their ability to actualize the unique qualities with which they have been endowed, and which make them different than all other creatures: rational structure […], complete centeredness, creative freedom, a possibility for self-actualization, and the ability for self-transcendence".[11] In short, being the *imago Dei* means possessing those qualities, those relationships and those functions similar (but not equal) to God's and to use them similarly (but not equally) as God would do. There are multiple causations. God created the world from within Godself and interacts with it from within the world through the nested hierarchy of the world. We can rewrite the equation as: The emergent universe (EU) comprises ultimate reality (UR), natural reality (NR) and emergent selves (ES) between which there is mutual

[11] http://www.pbs.org/faithandreason/theogloss/imago-body.html accessed 21 January 2016.

causation, but ultimate reality is bigger than both natural reality and emergent selves. Hence,

$$EU\left(UR \leftrightarrow NR \leftrightarrow ES\right); UR > \left(NR \cap ES\right).$$

I mentioned earlier that Zagzebski suggested that we should not understand God as a person; rather it is *personhood* that is the quality of humankind that distinguishes them from other instances of nature and that the personhood of human beings is the most significant way in which humans are like God. In my opinion, such a view sits very well with a panentheistic understanding of a divine relationship with the world and human beings. Accepting three-fold panentheism hence suggests that God possesses personhood. The 'en-relation' should, in my view, be understood as 'with-relation' in order to represent the relationship between God, the world and humans, i.e. God $\rightarrow$ (world $\cap$ humans). However, the panentheistic model suggested above implies that the world is the meeting place between God and human beings. Hence the 'en-relation' also denotes that God is 'with' human beings 'in' the world. Secondly, the 'en-relation' also needs to denote either an 'and-relation' or a 'with-relation' when it comes to the relationship from human beings toward God, because human beings have the *free will* to choose whether they want a 'with' or an 'and' relationship with God. A with-relationship from the human being toward God means that the human being 'wills' a friendship relationship with God. Because there always is a 'with' relationship from God towards human beings, whether the human beings recognize this or not; even if the human being does not want such a relationship, there will still be an 'and' relationship towards God. According to the Christian religion, God *calls* all human beings to be God's friends but does *not force* them. This implies that the 'and' relationship from human beings toward God may change into a 'with' relationship. Thirdly, the 'en-relation' will also denote an 'in-relation' between the world and God, i.e. the world is in God and God is in the world and human beings. However, since human beings are part of the world and part of God, they also need to be 'in' the world as well as 'in' God. To summarize, the *en* in pan*en*theism denotes several relationships:

1. With to denote the friendship relationship from God toward human beings
2. With or and to denote the friendship relationship from human beings toward God.
3. In to denote (a) the mutual relationship between God and the world and human beings, (b) the relationship between human beings and the world, i.e. human beings are in the world.

To conclude, there is only one emergent universe comprising both the natural and ultimate worlds forming a complex nested hierarchy in which parts are contained within wholes, which themselves become parts within greater wholes, etc. and there is a process of emergence. God or ultimate reality is seen as the ultimate ground of all necessities and possibilities and is in the natural world but is not reducible to it.

The world is God's similitude but human beings are also God's images; the world is the mediating authority where God and humans meet.

$$\mathrm{EU}(\mathrm{UR} \leftrightarrow \mathrm{NR} \leftrightarrow \mathrm{ES}); \mathrm{UR} > (\mathrm{NR} \cap \mathrm{ES}).$$

Conclusion

This chapter was dedicated to the question of how God or ultimate reality can be understood and how God or ultimate reality could act in the world. A first question was whether everyone has a relationship with God. The answer was that indeed, all have; even atheists take a stance in the sense of not God, but taking a stance is a way of relating. Relationships with God or ultimate reality are personal; i.e. they are subjective. Several models of understanding God were presented. Adapting classical theism, the most common view within Christianity, implies that God is put beyond time and space; this view sees God as a person and even a super-person, perfect in every way. The advantage of such a view is that it is easy to communicate with God, to talk and pray to God. After all, God is 'like us', only in a perfect manner. The disadvantages are many. However the problem tackled here was how an atemporal and aspacial God would be able to act in or intervene in a time-dependent and space-dependent world? The answer suggested was that God interacts with creation by way of miracles. While there are testimonies of people having experienced a miracle and while miracles are represented in the scriptures, there are obvious philosophical problems with such a view. One problem asks whether it is ethical to perform miracles for some and not for others. Furthermore, since miracles seem to occur randomly, how can that fit with the theistic all-loving God? Hence, there is the problem of theodicy. But there is also the problem of essence dualism; i.e. how could two such entirely different essences ever meet?

Another view was that God or ultimate reality constitutes only consciousness. The advantage of this view is that it does not suffer from the problem of essence dualism, since reality is seen as consciousness. The disadvantage is that material or earthly experiences are not important; these experiences belong to the world through which human beings simply have to pass. This has led to people seeing themselves as subjects of change rather than subjects who can make changes. The question about divine action in the world is irrelevant. Only one-way communication is possible, through meditation and contemplation, which may lead to a union with ultimate reality or God, but there is no interference or intervention from God or ultimate reality in life as such.

Another view presented was that God or ultimate reality is seen as emergence. This implies that the goal is to fulfil one's own destiny. This can be done by seeking harmony between contrasts within the world and oneself (e.g. good and bad). The advantage is that education, virtue and self-cultivation are highly appreciated. Also ultimate reality as emergence does recognize the importance of both ultimate and material life. But again, God or ultimate reality does not act in the world, similar to ultimate reality as consciousness; this view entails a personal one-way communication, from the person to God. Hence the question of how God or ultimate reality acts in the world is irrelevant.

The model of deism is similar to theism with the difference that while the theistic God intervenes in creation, the deist God created the world but once this was done, God chose to interfere no longer. The advantage of this view is that it fits very well with the theory of evolution. However, clearly, the question of divine action in the world is irrelevant.

The next model of God's relation to the world explored was pantheism. This view implies that God and creation are one; God is in the world and everything comprises an immanent God. The advantages are that there is no essence dualism, since God is literally present in everything. However, if God intervenes in creation, this can only be done from within.

However, despite the problems involved, some scholars wish to defend theism by way of enhancing some features of classical theism. One idea is to see God as being; not as 'ens' but as 'esse'. After all, Yahweh means 'I am' and not 'I am a super-person'; hence God is seen as being rather than a being. This line of thinking fits with the idea of God having personhood. A person has a body, but the concept of personhood need not imply embodiment. This again fits with the doctrine of the *imago Dei*, which is so essential for Christian theology. The idea is that humans are most like God in their personhood. Hence, empathy, friendship and love are placed as the focus. This idea fits well with the theistic doctrine of God as caritas.

The most important model for my purpose is panentheism. Panentheism dictates interdependence, and furthermore, there is no creation *ex nihilo*, but creation *continua*; i.e. God has not determined the cause of the universe; God interacts with the world and through this interaction changes the course of destiny. In other words, creation is still becoming. The advantages of this view are that the main features of theism are retained; e.g. God is still bigger than creation. However, there is no dualism because God is also inside the world, inside creation. God is both transcendent and immanent in relation to the world. Hence God's action in the world is both from within and without. However, the main question remains: How, more exactly, could God act in the world?

One suggestion came from Robert Cummings Neville who established the notion of the ontological creative act. The ontological creative act is indeterminate except in giving itself the nature of being creator of the world of determinate things. There is a kind of harmony between finite and infinite contrasts. This harmony is understood as things that fit together, which is different from a harmony between contrasts. These harmonies can be depicted in the world.

Another idea derives from Arthur Peacocke who argues that there is only one universe which comprises both natural and ultimate reality, in which through mutual affection of the hierarchies and the parts, new and distinct kinds of reality emerge. God is seen as the ultimate ground of all necessities and possibilities, which are two necessary features for a self-organized world. In other words, God is the ultimate ground for all that was, is and ever will be.

However, similar to Philip Clayton, the threefold panentheistic model I suggested includes the process of emergence. Furthermore, similar to Nicolaus Cusanus, my model embraces the doctrine of the *imago Dei*, embodied in the emergent self (ES), which consists of an objective neural self (ONS), a subjective neural self (SNS) and most importantly a subjective transcendent self (STS) which seems to be superior to the ONS and SNS. One important feature of the emergent self is personhood. An important feature of personhood is empathy. Because the emergent self is seen as the *imago Dei*, it follows that God has personhood and that divine caritas can also be understood as divine empathy. Hence, while the world is God's similitude, human beings, besides being God's similitudes, are also God's images and the world becomes the mediating authority where God and humans meet.

1. There is only one emergent universe (EU) comprising both ultimate reality (UR) and natural reality (NR).
2. God created the world from within Godself and interacts with it from within the world through the nested hierarchy of the world by way of multiple causations.
3. God is immanent and transcendent in relation to the world and human beings, the selves. $UR > (NR \cap ES)$.
4. The threefold panentheistic view includes the doctrine of the *imago Dei*; humans are both God's similitudes and *imago Dei*.
5. From this it follows that the world becomes the mediating authority between God and human beings.

I do not, of course, imply that this model can or even should be seen as any proof that God or ultimate reality exists, but at least it makes it harder to ignore or explain away the special feature of human beings in the way, for example, that evolutionary theorists attempt to do. Whether God acts in the world or not is a matter of faith and belief and since there are different belief systems and religious worldviews, the question of how God or ultimate reality acts in the world remains open.

Chapter 9
Final Conclusions and Reflections

Reaching the end of the present piece of research, it is time for some final conclusions and reflections. It is also time to propose further research because it is obvious that there is a lot more to say about what it is to be a human being. I have tried to answer this big question by making several distinctions. The first distinction was between *a (human) being – ens* and *(human) being – esse*. However, this distinction does not imply any dualism, because ens and esse, it was argued, are intertwined in a most complex manner. They are part of a whole, which I coined the *emergent threefold self*. Where the ens (called the objective neural self) is mostly related to the neuro-physiological underpinnings of being (human), esse is twofold. On the one hand there is a part of esse (called the *Subjective Neural Self*, SNS) that is mostly related to the neuro-physiological underpinnings (which I called the *Objective Neural Self*, ONS) and it is this part of esse that correlates with neuro-physiological activity. In other words, I suggested, there is a direct mutual causation between ens (ONS) and this part of esse (SNS). However, there seems to be another part of esse (called the *Subjective Transcendent Self*, STS) that, while also being in some correlation with ens, also has some correlation with the subjective neural self; furthermore, this part of esse seems to embrace both the objective and subjective neural self. In other words, it transcends the two. What I chose to call the subjective transcendent self has been given a variety of names in the history of philosophy. Even up to the present day, it has not been possible to actually account for it or actually grasp it. Nevertheless, this feature or process of the self seems to be very real. Perhaps it is this process of the self that is responsible for the way humans unreflectively understand themselves as dualist.

Perhaps some readers will see the model of the emergent threefold self as an analogy to the Christian model of human beings as composed of a body, a soul and a spirit. However, while one could think along those lines, there is a difference. Some also might see the present model as an analogy to the classical Trinitarian doctrine, which is the attempt to describe the Triune God and says that "(1) 'God is One' and (2) 'The one God is Trinity'" (Toom 2007: 10). Toom refers to Saint Augustine and writes "[t]he Father is God, and the Son is God, and the Holy Spirit

A.L.C. Runehov, *The Human Being, the World and God*,
DOI 10.1007/978-3-319-44392-8_9

is God, but the Son is not the Father, and the Holy Spirit is neither the Father nor the Son" (Toom 2007: 15).[1] The model of the emergent threefold self only concerns the human self and suggests that there is only one *closed* self in which, through the mutual causations of hierarchies, new distinct mental and neural properties emerge. The relationship between the three elements is as follows:

$$ES\left(ONS \leftrightarrow \left(SNS \cap STS\right); STS > \left(ONS \cap SNS\right)\right).$$

There is no essence dualism or essence trinitarianism. I found the distinction between ens and esse helpful for reflecting on the way a 'human being' might be assembled, but it was also fruitful to ponder about the purpose of human beings. Now, by purpose I do not mean the purpose of human life in the universe, which would imply human destiny. Related to this, I do not mean the purpose of the individual. Neither did I mean purpose in an evolutionary or religious sense. What I had in mind was to investigate some kind of purpose to (human) esse.

Hence, the question was whether there could be something special to (human) esse, whether there is a *sine qua non* about (human) esse. The first answer was that humans (and other species) cannot not experience. Hence experiencing became the sine qua non of esse. Sine qua non is not equal to uniqueness. Conditio sine qua non refers to an indispensable and essential trait, without which 'something' could not be. However, while all species experience, even flowers, it was argued, there is something special about human experiencing. Humans have the ability to analyze their experiences, which, as a matter of fact, is no more than another experience at another level of reality, a level of reality perhaps not (yet) attainable for other animals. Human experiencers want to understand why they experience. They want to know whether others experience as they do. They compare experiences. Human beings experience that they are experiencing; they understand the experiences of others. Humans have (even though not all) religious or spiritual experiences which they are able to account for (in their own terms). This line of thinking led me to further investigate whether, after all, there was something unique to human being.

The question was whether there may be something that is different from its being indispensable. The answer, it appears, lies in human social ontology. The level of social capacity is, of course, connected to the level of neural capacity in the species, and, needless to say, humans are far more developed neurologically compared to non-human societies, which gives them the opportunity to establish very advanced organizations. While all animals are in a way social animals (on different levels), only humans are institutional animals. In order to be a social animal, the animal has to possess three criteria. Firstly, the animal has to possess self– and group-identity as well as self-identity within a group. Secondly, the animal has to be able to

[1] I came to my attention that it would be interesting to discuss Trinitarian Theology in relation to my view of the threefold self. It is needless to say that I agree with this view. However, since I am not a theologian I dare not to put myself into this theological debate, at least not for the time being. I simply lack the time to do so. On the other hand, I warmly welcome theologians interested in the subject matter to continue and hence broaden the model I here suggested.

understand intentionality as well as collective intentionality. Thirdly, the animal needs to be able to commit to joint action. It was shown that humans are not unique in possessing these criteria; indeed; these criteria are met by the great apes and dolphins amongst other animals. What humans have established within their societies is, however, unique. Humans invented what Searle called deontic powers. Deontic powers perform a certain function not in virtue of physical structure but in virtue of collective attitudes. Examples of deontic powers are money, wedding rings, contracts, institutions, clock-time, and so forth. Some readers might see a connection between the human establishment of deontic powers and the religious doctrine of human beings as co-creators. In any case, humans do create for better or worse. The question of whether the cause of such creativity is due to co-evolution of genes and culture in a sense that these are mutualist, or due to the social complexity hypothesis which means that the more complex a society becomes, the more intelligence is acquired (leading to a continuing development of the human neo-cortex), remains unanswered and hence is open for further investigation. Human creativity does not end with establishing more and more advanced institutions or noetic powers. For some reason humans seem to strive to be able to copy themselves.

The topic of cloning was not discussed; instead the choice went to the creation of androids. The question raised was simply how far have computerscientists come in their quest to create human-like robots? In other words, how human are androids? It was argued that androids respond to each other, to humans and to human environments with a similar intelligence to humans, even though this intelligence is established by way of different features. However, it was maintained, that does not imply that contemporary androids also possess emotional intelligence. Up to the present day, androids do not have a type of theory of mind. They may recognize emotional symbols and may react to these symbols in an adequate manner but they do not do so on the basis of 'you are like me'. Androids may act compassionately, which is different from feeling compassionate. An android may show a sad face, which again is not the same as feeling sad. A robot may seem to be in love and behave like a human being in love but without feeling the butterflies in its stomach or thinking about the other person or android all the time. In any case, the philosophical question should no longer be 'Can a machine think?' but 'Can a machine feel?' In short, up to now, androids act as if they experience. Androids act on an external basis but still lack the internal basis, e.g. introspection.

The next big question tackled was what it is to be a human being in the world. What reality is and how it can be understood needed to be analyzed. Philosophically, this question belongs to the debate of realism. Not being entirely satisfied with the contemporary debate on realism, i.e. the variety of types of realism presented and defended, I suggested yet another type of realism which I called *extended realism.* Extended realism includes three dimensions in the way that human beings perceive reality. The first dimension concerns *measurable or observable reality*. The second dimension concerns *creational reality or reality* that is created by us; and the third dimension is about *the reality within us*, or for us, in other words, *phenomenological reality*. Measurable or observable reality is reality for us. Examples of such reality are black holes, rain, DNA, tsunami, stem-cells. This dimension of reality is intersubjective. Created reality, or reality from us for us, can be represented by, for example, currency, contracts, religion, wedding-rings (the noetic powers but also artifacts etc.) This dimension of reality is also intersubjective. The last type of reality, phenomenological reality, is reality within us for us. Within this reality we find, for example, love, sorrow, dreams, wishes, pain. This dimension is subjective and intersubjective. However, this dimension of reality also has a pure subjective dimension, a reality caused by brain disorders, narcotics, and so forth. All of reality is, in some way, reality for us. Since humans are part of reality, humans cannot know how reality actually is in itself. Humans do not have a 'divine point of view'. Humans are inside the box of reality. However, humans do have a fair understanding of reality if the following model is accepted. Humans live within (at least) three dimensions of reality, measurable, created and phenomenological reality.

When discussing possible models of reality, possible models of naturalism also need to be accounted for. Two new fruitful such models were suggested, *Minimalistic ontological naturalism* and *extended or flexible ontological naturalism.* A minimalist ontological naturalistic view of reality dictates that within a *specific realm* (R), if we have a mental state (P) then we have a physical/material state (Q); i.e. within a *specific realm* R: $\forall x P(x) \rightarrow Q(x)$, where P denotes mental states and Q all that is physical or material. However, it is necessary to note that the specific realm is related to one network of realities that includes both physical and mental reality and hence can be justified by way of reference to some part(s) in that network. *Extended or flexible ontological naturalism* states that there is only one natural order and that this comprises all of reality x. Nothing can be explained independently of the natural order. However, which theory, P(x) or Q(x), is the best one to explain the natural order is not determined. This type of naturalism can be formulated as $\forall x(P(x)$ v [or] $Q(x))$. This type of ontological naturalism seems to be more open than the former in that it takes account of the possibility of novel types of reality, mental as well as physical types of reality.

Another subject related to the esse of human being is the human capacity for mindreading. Mindreading is the capacity humans have to understand other minds. Once again, some distinctions had to be made. A first distinction was made between first- and third-person mindreading. It was argued that first-person mindreading also included mindwriting. Indeed, humans can read and write (re-write) their own mind. Following the classical theory of mindreading, it was argued that first-person mindwriting consists of reflecting about and adapting and/or changing a belief about (or metarepresenting) a psychological state of one's own. Concerning third-person mindreading and following the same theory, third-person mindreading consists of forming a belief (or judgment) about (or metarepresenting) another's psychological state. Obviously, mindwriting is not relevant for third-person mindreading. Mindwriting remains a capacity of the first-person perspective. Another distinction that was made and shown to be fruitful for discussing empathy was the distinction between low – and high level mindreading, suggested by Alvin Goldman. Hence, and since empathy is argued to be the core capacity of mindreading, a distinction was made between *low – and high – level empathy*. Low level empathy comprises the capacities of emotional contagion and biological altruism. These types of empathy are primitive, merely affective and mostly involuntary. High level empathy comprises the capacities of sympathy, selfless altruism and agape. These types of empathy are both affective and cognitive. The higher up the scale, the more cognitive they become. The bridge between low – and high level empathy is compassion, because compassion is still somewhat primitive, sometimes involuntary and sometimes highly affective, but it also has a cognitive component to be developed. From this reasoning, the potential of the mirror neuron theory to explain empathy was investigated. The conclusion drawn was that plain mirror neuron activity may explain low level empathy but not high level empathy because, in order to neurologically account for the activity involved in high level empathy, other neural functions are involved, not least the functions connected to cognition. Hence, high level empathy includes, besides mirroring, also simulation, imitation, imagination and hence theorizing. In short, when it comes to high level empathy the principle of analogical inference cannot be suspended.

The discussion on compassion took a different turn and showed the vulnerability of this mindreading capacity. Two views of compassion were compared, the anti-compassionate view and the pro-compassionate view. The advocates of the anti-compassionate view think that the *affective* component of compassion is irrational and should not be trusted as a guide to action. However, they also argue that compassion has a false *cognitive* structure. In other words, people do not think as they should do in relation to the other person. As mentioned above, compassion is a tricky issue because of its high affective level which is often primitive and involuntary. High affective levels may of course distort cognitive activity. The message given is, *you wipe your nose and I wipe mine,* (and we both keep our dignity). The pro-compassionate view, as the term makes clear, is more positive towards compassionate actions. However, there are restrictions. The message given, including the restrictions for compassionate action is, *if you are blameless, righteous and like me (i.e. belong to x social and ethnic community) then I will wipe your nose (i.e. I who am blameless, righteous and belong to s social and ethnic community).* Both views have their own specific philosophical problems. However, the problem they have in common is that they want to avoid the affective part of compassion. A solution to the problem is derived from Batson, who distinguished between empathy-induced altruism and moral motivation. Only when empathy-induced altruism (the affective component) and moral motivation (the cognitive component) cooperate, the Golden Rule is applied. In addition, it was argued that besides neurological preconditions (having the ToM and no neural deficiencies), factors such as culture, religion, education and politics seem to want to have a voice as well. The tricky question that remains is this: Is an action that looks like a compassionate action always a true compassionate action, or can there be other motivations involved other than wanting to give a helping hand?

When discussing empathy and compassion one also needs to consider the evil in the world as well as the question of whether humans possess free will and to what extent. The latter question is related to the extent of responsibility humans have for their actions. The investigation began by considering two views of responsibility. On the one hand, there are libertarianists who argue that one is always responsible for one's actions; on the other hand, incompatibilists who argue that such cannot be the case due to the fact that humans are dependent on their environment (culture, education, religion and so forth). Both views were shown to have some philosophical problems. The libertarianists need to explain in what sense causally undetermined free acts really are free, while the incompatibilists need to explain in what way humans could be held responsible for their acts presupposing everything is predetermined. It was argued that libertarianism cannot account for accidental non-intentional moral evil acts, while incompatibilism cannot account for active intentional moral evil acts. In order to try to solve the problem, four types of moral evil were suggested; namely, *pure accidental moral evil, belief-based moral evil, active but not entirely responsible moral evil and active moral evil.* A middle road between libertarianism and incompatibilism was taken. Indeed, people should not always be considered entirely responsible for their deeds, but they should be held responsible in some cases, and sometimes the circumstances are such that they cannot be held responsible at all. The importance of life, and worldviews in decision making (for better or worse) was also discussed. The importance of neuroscience, nanotechnology and genetics for minimizing evil actions was considered. However, this led to the suggestion that such research, not least when put into praxis should be strongly guided by ethical and juridical rules in order to protect human rights.

The final discussion on the human esse in the world concerned what time is and how it can be understood. There seems to be nothing more mysterious that the notion of time. Time was divided into time as studied (external time) and time as experienced (internal time). Time, as studied, included Newtonian, Einsteinian and quantum understandings of time. Newton's time is linear, presented as an arrow from the past, moving over the present towards the future. Newton's view of time is static and problematic for several reasons. One reason is that time cannot be seen as an arrow at all; in fact, there are several arrows on a timeline that may point in different directions, as Boltzmann argues. Boltzmann's view of time is dynamic, something time actually is. Another reason Newton's view is problematic is that time and space are not distinct as he believed. Hence, Einstein presented the idea of space-time and made a distinction between coordinate and proper time. Coordinate time is only subjective in relation to the observer. Proper time is the same for all observers. While there are several interpretations of time within quantum physics, one is more dominant, namely that time does not exist before a measurement or observation takes place. The consequence is that time, quantum mechanically speaking, is always 'now'. Scientific understandings of time recognize no flow of time such as that experienced by humans. Hence, experienced time had to be investigated.

There are at least three main types of experienced time. Firstly, time is experienced as a circle. Perhaps cyclic time, the wheel of time, or Kalachakra, tells us more about transformation than time. Seasons follow each other year after year; everything comes into life only to disappear from life again, in an endless cosmic circle. Secondly, be it very rarely, time is not experienced at all. The example given was the Pirahã Tribe of the Amazon rainforest. These people do not have a notion of time. However, this is due to their very limited language. Hence, the notion of time seems to be connected to language. Thirdly, and the most common, is that time is experienced as a flow. This flow is then understood as transience, a kind of moving on, or as a succession, i.e. as one event after the other. Several explanations were presented for what the experience of time as flow can be due to, but none was entirely satisfactory. It was argued that the experience of time is sometimes (if not always) relative to the experiencer. Sometimes one experiences time as moving very slowly; for example, when one is bored or longs for something; and sometimes one experiences time as moving too fast, for example when one is very busy or on holiday. Sometimes one has the feeling that time has stopped. Hence, some aspects of time are very personal. Therefore it was suggested that people have an inner personal clock. On the other hand, the experience of time is also intersubjective. This aspect of time, I argued, can be understood as a deontic power. Indeed, this aspect of time is invented or created by us. Indeed, in order to have an international civil time standard, Greenwich Mean Time (GMT) was created, which was later superseded by the Coordinated Universal Time (UTC). Hence, even though humans experience time in a personal manner, it does not imply that time is relative, thanks to human creativity and deontic powers.

The last part concerned possible relationships between God or ultimate reality, human beings and the world. The main question was how God or ultimate reality could act in the world. It was argued that all, including atheists, have a relationship with God or ultimate reality. Atheists who argue for not-God take a stance and taking a stance is a way of relating. Several models of God or ultimate reality were presented. It was shown that there are three models of God or ultimate reality which are irrelevant for discussing possible divine action in the world. These models were: God as pure consciousness; God as emergence; and Deism. The model of pantheism gives room for divine action in the world; however, in this case divine action comes from within the world, since God is equal to the world and is immanent and transcendent in all there is. The model of theism seemed to be the most problematic for understanding possible divine action in the world perhaps because the most common understanding within theism is that God is entirely different from creation. God is space-less and timeless. Hence, the question of how a space- and timeless God could act in the world is a most challenging one. One proposed solution derives from how time is understood within quantum physics. If this view is accepted, it was argued, God does not need to be in time; it is enough that the observer is. Nevertheless, this suggestion is not without problems. One particular problem is that God could not intervene in the world if not asked to intervene. It was argued that panentheism provides the most fruitful option for possible divine action in the world. Inspired by Arthur Peacock, Cusanus, Philip Clayton, Zagzebski and Celia Deane-Drummond, the following threefold panentheistic model was suggested.

1. There is only one emergent universe (EU) comprising both ultimate reality (UR) and natural reality (NR).
2. God created the world from within Godself and interacts with it from within the world through the nested hierarchy of the world by way of multiple causations.

3. God (UR) is immanent and transcendent in relation to the world (NR) and human beings (ES), emergent selves. UR > (NR ∩ ES).
4. The threefold panentheistic view includes the doctrine of the imago Dei; humans are both God's similitudes and imago Dei.
5. From this it follows that the world becomes the mediating authority between God and human beings.

However, even if this model is promising, the question of whether God acts in the world or not is a matter of faith and belief, and since there are different belief systems or religious worldviews, the question of how God or ultimate reality acts in the world cannot be settled.

Studying human being is a very complex subject matter of study. I like the end this investigation with the words of Philip Clayton with whom I agree.

> You can 'wager' that the 'real' things that exist in the world are physical processes within organisms, and that everything else – intentions, free will, ideas like justice or the divine – are "constructs", complicated manifestations of neural processes. But I am wagering on the other side. I wager that no level of explanation short of irreducibly psychological explanations will finally do an adequate job of accounting for the human person [human being]. And this means, [...] the real existence and causal efficacy of the conscious or the mental dimension [esse] of human personhood [being]. (Clayton 1999: 211).

Bibliography

Aargaard, M. 2007. *Berglinske*, Thursday 8 November, 09:02.

Adams, William A. 2006. Transpersonal heterophenomenology. *Journal of Consciousness Studies* 13(4): 89–93.

Adolphs, Ralph, Daniel Tranel, Hanna Damasio, and Antonio R. Damasio. 1995. Fear and the human Amygdala. *Journal of Neuroscience* 15: 5879–5891.

Aharoni, E., W. Sinntt-Armstrong, and K.A. Keihl. 2012. Can psychopathic offenders discern moral wrongs? A neo look at the moral/conventional distinction. *Journal of Abnormal Psychology* 121: 484–497.

Alexander, Samuel. 1920. *Space, time and deity*. London: Macmillan.

Allee, Walter Clyde. 1931. Co-operation among animals. *The American Journal of Sociology* 37(3): 386–398.

Alston, William P. 1991. *Perceiving god: The epistemology of religious experience*. New York: Cornell University Press.

Anderberg, Thomas. 1997. *Guds moral: En essä om lidandets och ondskans problem*. Nora: Nya Doxa.

Aniskiewicz, Albert S. 1979. Autonomic component of vicarious conditioning and psychopathy. *Journal of Clinical Psychology* 35: 60–67.

Anselm of Canterbury. 2008. *Anselm of Canterburry: The major works*. Oxford: Oxford University Press.

Aquinas, Thomas. 1991. *Summa theologiæ, a concise translation,* ed. Timothy McDermott. Westminister: Christian Classics.

Arendt, Hannah. 2006. *Eichmann in Jerusalem. A report on the Babality of evil*. New York: Penguin Classics.

Aristotle. 1999. *Nicomachean ethics*. Indianapolis: Hacked Publishing Co. Inc.

Aristotle. 2005. *On the soul.* Trans. John Alexander Smith. Stilwell: Digireads.com Publishing.

Arkin, R.C. 2005. Moving up the food chain: Motivation and emotion in behaviour-based robots. In *Who needs emotions: The brain meets the robot,* eds. J. Fellous and M. Arbib. Oxford: Oxford University Press. http://www.cc.gatech.edu/ai/robot-lab/online-publications/moral-final2.pdf

Audi, Robert (ed.). 1999. *The Cambridge dictionary of philosophy*, 2nd ed. Cambridge/New York/Melbourne/Madrid: Cambridge University Press.

Augustine. 2011. *The complete works,* ed. Philip Schaff, Trans. Marcus Dods and Rose Elisabeth. Cleveland: Kindle edition: ASIN B005FY66EA.

Barbour, Ian. 1999. Neuroscience, artificial intelligence and human nature: Theological and philosophical reflections. In *Neuroscience and the person: Scientific perspectives on divine action,*

A.L.C. Runehov, *The Human Being, the World and God,*
DOI 10.1007/978-3-319-44392-8

ed. R.J. Russel, N. Murphy, T.C. Meyering, and M.A. Arbib, 215–248. Vaticcan City State: Vatican Observatory Publications.

Barbour, Julian. 2000. *The end of time, the next revolution in our understanding of the universe.* London: Phoenix, Orion Books, Ltd.

Barrett, Justin L. 2004. *Why would anyone believe in god?* Walnut Creek: AltaMira Press.

Batson, Daniel C. 2014. Are the 'principles in his nature, which interest him in the fortune of others', moral? In *Understanding moral sentiments. Darwinian perspectives?* ed. Hilary Putnam, Susan Neiman, and Jeffrey P. Schloss, 43–58. London/New Brunswick: Translation Publishers.

Beier, Matthias. 2006. On the psychology of violent christian fundamentalism: Fighting to matter ultimately. *Psychoanalytic Review* 93(2): 301–327.

Bergson, Henri. 1965. *Duration and simultaneity.* Indianapolis/New York/Kansas City: The Bobbs-Merrill Company, Inc.

Boethius. 1999. *The consolation of philosophy.* London/New York: Penguin Books.

Bohm, David, and Basil J. Hiley. 1993. *The undivided Universe.* New York: Routledge.

Bohr, Hans. 1967. My father. In *Niels Bohr: His life and work as seen by his friends*, ed. Rozental Stefan, 328. Amsterdam: Elsevier/North-Holland Pub. Co.

Boltzmann, Ludwig. 1895. On certain questions of the theory of gases. *Nature* 51: 413.

Bonaventura. 1990. *The journey of the mind to god.* Trans. Philotheus Boehner, ed. Stephen F. Brown. Indianapolis/Cambridge: Hackett Publishing Co.

Bowlby, John. 1951. Maternal care and mental health. *Bulletin of the World Health Organization*, Geneva: WHO.

Boydston, Jo Ann., ed. 1967–1991. *The collected works of John Dewey.* (37 Vols.), Carbondale: Southern Illinois University Press.

Brass, Marcel, Ruth M. Schmitt, Stephanie Spengler, and György Gergely. 2007. Investigating action understanding: Inferential processes versus action simulation. *Current Biology* 17: 2117–2121.

Brentano, Franz. 2015. *Psychology from an empirical standpoint.* Foreword by Tim Crane. London/New York: Routledge.

Bretherton, Inge, J. Fritz, C. Zahn-Waxler, and D. Ridgeway. 1986. The acquisition and development of emotion language: A functionalist perspective. *Child Development* 57: 529–548.

Brooke, John Hedley. 2006. Science and the self: What difference did Darwin make? In *The evolution of rationality*, ed. F.L. Shults, 253–277. Cambridge: Eerdmans Publishing Co.

Brothers, Lesley. 1989. A biological perspective on empathy. *American Journal of Psychiatry* 146: 10–19.

Browne, Derek. 2004. Do Dolphins know their own minds? *Biology and Philosophy* 19: 633–653.

Bultmann, Rudolf. 1951–1955. *Theology of the new testament.* (2 Volumes). New York: Scribner.

Burns, Charlene P.E. 2002. *Divine becoming: Rethinking Jesus and incarnation.* Minneapolis: Ausburg Fortress Press.

Byers, Sarah. 2012. The psychology of compassion: Stoicism in city of god 9.5. In *The Cambridge critical guide to the city of god*, ed. James Wetzel. Cambridge: Cambridge University Press.

Byrne, Alex, and David R. Hilbert. 2003. Color realism and color science. *The Behavioral and Brain Sciences* 26: 3–64.

Byrne, Richard, and Andrew Whiten (eds.). 1988. *Machiavellian intelligence: Social expertise and the evolution of intellect in monkeys, apes and humans.* Oxford: Oxford University Press.

Calaprice, Alice. 2005. *The new quotable Einstein.* Princeton/Oxford: Princeton University Press/ Hebrew University of Jerusalem.

Castro-Cladas, Alexander, Karl Magnus Petersson, A. Reis, Sharon Stone-Elander, and Martin Ingvar. 1998. The illiterate brain learning to read and write during childhood influences the functional organization of the adult brain. *Brain* 121: 1053–1063.

Churchland, Patricia. 2002. *Brain-wise.* Cambridge/London: MIT Press.

Clayton, Philip. 1999. Neuroscience, the person, and god: An emergentist account. In *Neuroscience and the person: Scientific perspectives on divine action*, ed. R.J. Russel, N. Murphy,

T.C. Meyering, and M.A. Arbib, 181–214. Vaticcan City State: Vatican Observatory Publications.
Clayton, Philip. 2004. *Mind & emergence, from quantum to consciousness*. Oxford: Oxford University Press.
Clayton, Philip, and Arthur Peacocke (eds.). 2004. *In whom we live and move and have our being*. Cambridge: Eerdmans Publishing Co.
Conway, Anne. 1996. *The principles of the most ancient and modern philosophy*. New York: Cambridge University Press.
Cupitt, Don. 2002. *Is nothing sacred? The Non-realist philosophy of religion: Selected essays*. New York: Fordham University Press.
Curd, Patricia, and Daniel W. Graham (eds.). 2008. *The Oxford handbook of presocratic philosophy*. Oxford: Oxford University Press.
d'Aquili, Eugene G., and Andrew B. Newberg. 1999. *The mystical mind: Probing the biology of religious experience*. Minneapolis: Fortress Press.
Darwin, Charles. 2009 [1874]. *The descent of man, and selection in relation to sex*. Digireads.com publication. www.digireads.com
Davies, Oliver. 2001. *A theology of compassion*. Cambridge: SCM-Canterbury Press.
Dawkins, Richard. 2001. How has the world changed? *The Guardian*. Interview
Dawkins, Richard. 2006. *The god delusion*. London: Transworld Publishers.
Deane-Drummond, Celia. 2004. The logos as wisdom: A starting point for a sophianic theology of creation. In *In whom we live and move and have our being*, ed. Philip Clayton and Arthur Peacocke, 233–248. Cambridge: Eerdmans Publishing Co.
Decety, Jean. 2010. The neurodevelopment of empathy in humans. *Developmental Neuroscience* 32: 257–267.
Decety, Jean. 2014. The neuroevolution of empathy and caring for others: Why it matter for morality. In *New frontiers in social neuroscience*, ed. Jean Decety and Yves Christen, 127–152. Heidenberg/New York/Dordrecht/London: Springer.
Decety, Jean, and Philip L. Jackson. 2007. The functional architecture of human empathy. *Behavioral and Cognitive Neuroscience Reviews* 3(2): 71–100.
Decety, Jean, and J.A. Sommerville. 2003. Shared representations between self and other: A social cognitive neuroscientific view. *Trend in Cognitive Science* 7: 527–533.
Dennett, Daniel. 1987. *Intentional systems in cognitive ethology. The intentional stance*. Cambridge: MIT Press.
Dennett, Daniel. 1992. The self as a center of narrative gravity. In *Self and consciousness, multiple perspectives*, ed. F. Kessel, P. Cole, and D. Johnson. Hillsdale: Erlbaum.
Dennett, Daniel. 1993. *Consciousness explained*. London: Penguin Books.
Derrida, Jacques. 1982. *Margins of philosophy*. Trans. A. Bass. Chicago: Chicago University Press.
di Pellegrino, G., L. Fadiga, L. Fogassi, V. Gallese, and G. Rizzolatti. 1992. Understanding motor events: A neurophysiological study. *Experimental Brain Research* 91: 176–180.
Diller, Jeanine, and Asa Kasher. 2013. *Models of god and alternative ultimate realities*. Dordrecht/ Heidenberg/New York/London: Springer.
Diniz, Breno S., Meryl A. Butters, Steven M. Albert, Mary Amanda Dew, and Charles F. Reynolds. 2013. Late-life depression and risk of vascular dementia and Alzheimer's disease: Systematic review and meta-analysis of community-based cohort studies. *The British Journal of Psychiatry* 202(5): 329–335.
Drees, Willem B. 1998. *Religion, science, and naturalism*. Cambridge: Cambridge University Press.
Drees, Willem B. 2003. Naturalism. In *The encyclopedia of science and religion*, ed. Huyssteen Wentzel Van, 593–597. New York: Macmillan Reference.
Drees, Willem B. 2010. *Religion and science in context*. Abingdon: Routledge.
Dretske, Fred I. 1999. Machines, plants and animals: The origin of agency. *Erkenntnis* 51: 19–31.

Dunbar, Robin I.M. 1995. Neocortex size and group size in primates: A test of the hypothesis. *Journal of Human Evolution* 28(3): 287–296.

Eccles, John C., and Karl R. Popper. 1981. *The self and its brain*. Berlin/Heidelberg/London/New York: Springer.

Egner, Robert E., and Lester E. Denonn (eds.). 2009. *The basic writing of Bertrand Russell*. London/New York: Routledge.

Espinas, Alfred Victor. 1878. *Des Sociétés animals*. Paris: Germer Baillière.

Farrer, Chloe, N. Franck, N. Gerogieff, C.D. Frith, J. Decety, and M. Jeannerod. 2003. Modulating the experience of agency: A positron emission tomography study. *NeuroImage* 18: 324–333.

Feinberg, Todd E., and Julian Paul Keenan. 2005. Where in the brain is the self? *Consiousness and Cognition* 14: 661–678.

Fieldman, George. 2003. Evolution and evil. *Perspectives on Evil and Human Wickedness* 1(3): 122–132.

Flanagan, Owen. 1992. *Consciousness reconsidered*. Cambridge: MIT Press.

Flanagan, Owen. 2006. Varieties of naturalism. In *The Oxford handbook of religion and science*, ed. P. Clayton and Z. Simpson, 430–452. Oxford: Oxford University Press.

Flood, Raymond, and Michael Lockwood. 1986. *The nature of time*. Oxford: Basil Blackwell Ltd.

Fossati, P., S.J. Hevenor, S.J. Graham, C. Grady, M.L. Keightley, F. Craik, and H. Mayberg. 2003. In search of the emotional self: An fMRI study using positive and negative emotional words. *American Journal of Psychiatry* 160(11): 1938–1945.

Frankenhaeuser, Marianne. 1959. *Estimation of time, an experimental study*. Stockholm: Almqvist & Wiksell.

Freud, Sigmund. 1961. *Civilization and its discontent*. New York: WW Norton & Co., Inc.

Gabrielsson, Johannes. 1926. *Augustinus*. Stockholm: A. Bonniers Förlag.

Gallese, Vittorio. 2001. The 'shared manifold' hypothesis: From mirror neurons to empathy. *Journal of Consciousness Studies* 8(5–7): 33–50.

Gallese, Vittorio. 2003. The roots of empathy: The shared manifold hypothesis and the neural basis of intersubjectivity. *Psychopathology* 23: 171–180.

Gallese, Vittorio. 2005. Embodied simulation: From neurons to phenomenal experience. *Phenomenology and the Cognitive Sciences* 4: 23–48.

Gallese, Vittorio, Luciano Fadiga, Leonardo Fogassi, and Giacomo Rizzolatti. 1996. Action recognition in the premotor cortex. *Brain* 119: 593–609.

Gallese, Vittorio, Christian Keysers, and Giacomo Rizzolatti. 2004. A unifying view of social cognition. *Trends in Cognitive Science* 8: 396–403.

Gazzaniga, Michael S. 2008. *The split brain revisited. Groundbreaking work that began more than a quarter of a century ago has led to ongoing insights about brain organization and consciousness.* http://www.utdallas.edu/~otoole/CGS2301_S09/7_split_brain.pdf. Accessed 25 Jan 2016.

Gill, Mary Louise, and Pierre Pellegrin (eds.). 2006. *A companion to ancient philosophy*. Malden/Oxford: Wiley-Blackwell.

Glendinning, Simon. 2004. Language. In *Understanding Derida*, ed. J. Reynolds and J. Roffe, 5–13. New York: Continuum.

Goetz, Stewart, and Charles Taliaferro. 2008. *Naturalism*. Cambridge: Wim B. Eerdmans Publising Co.

Goldberg, Elkhonon. 2001. *The executive brain, frontal lobes and the civilized mind*. Oxford/New York: Oxford University Press.

Goldman, Alvin. 1999. Naturalism. In *The Cambridge dictionary of philosophy*, ed. R. Audi, 596–599. Cambridge: Cambridge University Press.

Goldman, Alvin. 2000. What is justified belief. In *Epistemology. An antology*, ed. E. Sosa and K. Jaegwon, 89–104. Maldern/Oxford: Blackwell Publishers.

Goldman, Alvin. 2006. *Simulating minds, the philosophy, psychology and neuroscience of mind-reading*. Oxford/New York: Oxford University Press.

Goleman, Daniel. 1997. *Emotional intelligence, why it can matter more than IQ*. New York: Bantam Books.
Gordon, Robert. 2005. Chapter 15: Intentional agents like myself. In *Perspectives on imitation: From neuroscience to social science*, ed. Susan Hurley and Nick Chater. Cambridge: MIT Press.
Green, Joel B. 2008. *Body, soul, and human life*. Grand Rapids: Baker Academic.
Gregersen, Niels Henrik. 2004. Three varieties of Panentheism. In *In whom we live and move and have our being*, ed. Philip Clayton and Arthur Peacocke, 19–35. Cambridge: Eerdmans Publishing Co.
Gregersen, Niels Henrik. 2008. *Levels of complexity*. www.metanexus.net. Accessed 27 May 2015.
Griffin, David Ray. 2000. *Reenchantment without supernaturalism: A process philosophy of religion (Cornell studies in the philosophy of religion)*. New York: Cornell University Press.
Guttenplan, Samuel. 2001. *A companion of the philosophy of mind*. Oxford: Blackwell.
Harms, Ernest. 1944. The development of religious experience in children. *American Journal of Sociology* 50: 12–22.
Hançil, Tomas, and M. Ziemer. 2000. Evolutionary epistemology as a new challenge in the dialogue between theology and science. *CTNS Bulletin* 20(3).
Hanisch, Helmut. 1996. Gottesbilder. Eine empirische Untersuchung bei religiös und nicht-religiös erzogenen Kindern und Jugendlichen im Alter von 7–16. In: *Christenlehre-PraxisReligionsunterricht*, 49. Jg., 2: 4–11.
Harris, S.M. 2006. *The end of faith*. London: Simon & Schuster.
Heidegger, Martin. 1993. *Sein und Zeit*. Tübingen: Max Niemeyer Verlag.
Helm, Paul. 1988. *Eternal god, a study of god without time*. Oxford: Clarendon Press.
Herrmann, Eberhard. 2004. *Religion, reality and a good life*. Tübingen: Mohr Siebeck.
Herrmann, Eberhard. 2008. On the distinction between the concept of god and conceptions of god. *International Journal for Philosophy of Religion* 64: 63–73.
Hickok, Gregory. 2014. *The myth of mirror neurons*. http://www.worldsciencefestival.com/2014/08/smart-reads-gregory-hickoks-myth-mirror-neurons/. Accessed 2 Feb 2016.
Hinterberger, Thilo. 2011. *Folrasonium – Music played by plants*. http://www.florasonium.de/english_site.html. Accessed 11 Dec 2015.
Hobbes, Thomas. 1654. *Treatise of 'Liberty and Necessity'*. Available on http://isites.harvard.edu/fs/docs/icb.topic1223571.files/hobbes%20on%20liberty.pdf. Assessed 8 Feb 2016.
Höffe, Otfried. 2000. *Immanuel Kant*. Stockholm: Thales.
Holte, R. 1984. *Människa, Livstolkning, Gudstro*: Teorier och metoder inom tros- och livsåskådningsvetenskapen. Bodafors: Doxa.
Hook, Sidney. 1961. *The quest for being, and other studies in naturalism and humanism*. Westport: Greenwood Press.
Hopkins, D. 1997. Failing brain, faithful community. In *God never forgets*, ed. Donald K. McKim, 72–87. Louisville: John Knox Press.
Hume, David. 1978. In *A treatise of human nature*, ed. L.A. Selby-Bigge. Oxford: Clarendon Press.
Hume, David. 2008. *Enquiry concerning human understanding*. Oxford: Oxford University Press.
Hunt, Matthew O. 1996. The individual, social, or both? A comparison of black, latina, and white beliefs about the causes of poverty. *Oxford Journals, Social Forces* 75(1): 293–322.
Husserl, Edmund. 1984. *Logische Untersuchungen. Zweiter Band: Untersuchungen zur Phänomenolgie und Theorie der Erkenntnis*. Den Haag: Nijhoff.
Hyman, Gavin. 2010. *A short history of atheism*. London: I.B. Tauris.
Iacoboni, Marco. 2009. Imitation, empathy and mirror neurons. *Annual Review of Psychology* 60: 653–670.
Ishiguro, Hiroshi. 2005. Android science: Toward a new cross-interdisciplinary framework. *Cognitive Science Society* 25–26 (July): 1–6.
Jacob, Pierre. 2009. A philosopher's reflections on the discovery of mirror neurons. *Topics in Cognitive Science* 1: 570–595.

Jacobs, Jon. 2008. Naturalism. *Internet encyclopedia of philosophy*. http://www.iep.utm.edu/naturali/. Accessed 2 Feb 2016.

Johnson, Sterling C.J., Leslie C. Baxter, Lana S. Wilder, James G. Pipe, Joseph E. Heiserman, and George P. Prigatano. 2002. Neural correlates of self-reflection. *Brain* 125: 1808–1814.

Jonas, Hans. 2001. *The gnostic religion*. Boston: Beacon Press.

Jones, Kathleen, B. 2014. The trial of Hannah Arendt. *Humanities* 35(2). http://www.neh.gov/humanities/2014/marchapril/feature/the-trial-hannah-arendt. Accessed 21 Aug 2014.

Kant, Immanuel. 1960. *Religion within the limits of reason alone*. New York: The Open Court Publishing Co.

Kant, Immanuel. [1781] 2004. *Kritik av det rena förnuftet*. Stockholm: Thales.

Kant, Immanuel. 2014. http://plato.stanford.edu/entries/kant-religion/. Accessed 5 Feb 2016.

Keck, David. 1996. *Forgetting whom we are: Alzheimer's disease and the love of god*. Nashville: Abingdon Press.

Keyser, Christian, and David I. Perrett. 2004. Demystifying social cognition: A Hebbian perspective. *Trends in Cognitive Sciences* 8: 501–507.

Klunk, William, E. 2003. New imaging agent detects amyloid plagues in Alzheimer disease. *Neurology Today*, 3(12):1, 15–16.

Kohls, Niko, and Harald Walach. 2006. Exceptional experiences and spiritual practice: A new measurement approach. *Spiritual and Health International* 7: 125–150.

Konstantinov, I.E. 2000. In search of Alexander A. Maximow: The man behind the unitarian theory of hematopoiesis. *Perspectives in Biology and Medicine* 43(2): 269–276.

Levinas, Emanuel. 1969. *Totality and infinity*. Trans. A. Lingis. The Hague: Marinus Nyhoff.

Lockwood, Michael. 1989. *The mind, the brain and the quantum*. Cambridge: Basil Blackwell Ltd.

MacFarlane, Alistair. 2013. Information, knowledge & intelligence. *Philosophy now: A magazine of ideas*. https://philosophynow.org/issues/98/Information_Knowledge_and_Intelligence. Accessed 1 Feb 2016.

Machiavelli, Niccoló. 1998. *The Prince*. Trans. and intro. C. Harvey. Mansfield/Chicago/London: Chicago University Press.

Maulana, Mohammed Razi Khan Afridi. 2011. *Manual of Islam: Islamic Sharia on faith*, vol. 2. Berkeley/Los Angeles/London: Pentagon Press.

McCarthy, John. 1979. Ascribing mental qualities to machines. In *Philosophical perspectives in artificial intelligence*, ed. M. Ringle. New Jersey: Atlantic Highlands.

McCarthy, John and Patrick J. Hayes. 1969. *Some philosophical problems from the standpoint of artificial intelligence*. http://www-formal.stanford.edu/jmc/mcchay69.pdf, 1–51. Accessed 1 Feb 2016.

McGrath, Alister. 2007. *The Dawkins Delusion*. London: Society for Promoting Christian Knowledge.

Meltzoff, Andrew N. 2005. Imitation and other minds: The "like me" hypothesis. In *Perspectives on imitation: From neuroscience to social science*, ed. Susan Hurley and Nick Chater, 55–77. Cambridge: MIT Press.

Meltzoff, Andrew N. 2007. 'Like me': A foundation for social cognition. *Developmental Science* 10(1): 126–134.

Meltzoff, Andrew N., and Keith Michael Moore. 1983. Newborn infants imitate adult facial gestures. *Child Development* 54: 702–709.

More, E. *Basilides of Alexandria*. http://www.iep.utm.edu/gnostic/print. Accessed 18 Jan 2016.

Morse, M., P. Castillo, D. Venecia, J. Milstein, and D.C. Tyler. 1986. Childhood near-death experiences. *American Journal of Diseases of Children* 140(11): 299–318.

Mother Teresa. 2007. *Come be my light, the revealing private writings of the Nobel Peace Prize Winner*. ed. and com. Brian Kolodiejchk. New York: Doubleday.

Murphy, Nancey. 1998. Supervenience and the nonreducibility of ethics to biology. In *Evolutionary and molecular biology: Scientific perspectives on divine action*, ed. Robert John Russell, R. William, S.J. Stoeger, and Francisco J. Ayala, 463–489. Pasadena: Vatican Observatory.

Naess, Arne. 1993. *Beautiful action. Its function in the ecological crisis.* Environmental values. Cambridge: The White Horse Press, 67–71.

Nagel, Thomas. 1986. *Utsikten från ingenstans.* Nora: Nya Doxa.

Nelkin, Norton. 2001. Subjectivity. In *A companion to the philosophy of mind*, ed. Samuel Guttenplan. Oxford: Blackwell Publishers.

Neville, Robert Cummings. 2013. Modeling ultimate reality: God, consciousness and emergence. In *Models of god and alternative ultimate realities*, ed. Jeanine Diller and Asa Kasher. Dordrecht: Springer.

Newberg, Andrew B., and Mark Robert Waldman. 2009. *How god changes your brain.* New York: Ballantine Books.

Newberg, Andrew B., Abass Alavi, Michael Baime, Michael Pourdehnad, Jill Santanna, and Eugene G. d'Aquili. 2001. The measurement of reginal cerebral blood flow during the complex cognitive task of meditation: A preliminary SPECT study. *Psychiatry Research Neuroimaging Section* 106: 113–122.

Nishitani, Keiji, and Robert E. Carter. 2006. *On Buddhism.* New York: State University of New York Press.

Nussbaum, Martha C. 2008. *Upheavals of thought, the intelligence of emotion.* New York/Melbourne/Madrid/Cape Town/Singapore/Säo Oaulo/Delhi: Cambridge University Press.

Nussbaum, Martha C. 2014. Compassion: Human and animal. In *Understanding moral sentiments. Darwinian perspectives?* ed. Hilary Putnam, Susan Neiman, and Jeffrey P. Schloss, 123–150. London/New Brunswick: Translation Publishers.

Ochsner, Kevin N., Kyle Knierim, David H. Ludlow, Josh Hanelin, Tara Ramachandran, Gary Glover, and Sean C. Mackey. 2004. Reflecting upon feelings: An fMRI study of neural systems supporting the attribution of emotion to self and other. *Journal of Cognitive Neuroscience* 16(20): 1746–1772.

Okasha, Samir. 2008. Biological Altruism. *Stanford encyclopedia of philosophy* http://stanford.library.usyd.edu.au/archives/spr2008/entries/altruism-biological/. Accessed 26 Jan 2016.

Oord, Thomas Jay. 2015. *The uncontrolling love of god.* Downers Grove: IVP Academic Press.

Otto, R. 1970. *The idea of the holy.* London: Oxford University Press.

Overall, Christine. 2003. Miracles and Larmer. *Dialogue* 42(1): 123, 355–366.

Overbye, Dennis. 1991. *Lonely hearts of the cosmos: The scientific quest for the secret of the universe.* New York: HarperCollins.

Oviedo, Lluis. 2006. Religious experience: First-, second-, and third-person accounts. *Archivio Di Filosofia* 13: 391–401.

Palmer, Stephen E. 1999. *Vision science: Photons to phenomenology.* Cambridge, MA/London: MIT Press.

Panksepp, Jaak. 1998. *Affective neuroscience: The foundations of human and animal emotions.* London: Oxford University Press.

Papineau, David. 2007. Naturalism. *Stanford encyclopedia of philosophy.* http://plato.stanford.edu/entries/naturalism/. Accessed 2 Feb 2016.

Peacocke, Arthur. 2002. *Paths from science towards god.* Oxford: Oneworld Publications.

Peacocke, Arthur. 2004. Articulating god's presence in and to the world unveiled by the sciences. In *In whom we live and move and have our being*, ed. Philip Clayton and Arthur Peacocke, 137–154. Cambridge: Eerdmans Publishing Co.

Persinger, Michael A. 1987. *Neuropsychological bases of god beliefs.* New York: Praeger.

Peters, Ted. 2013. Models of God. In *Models of god and alternative ultimate realities*, ed. Jeanine Diller and Asa Kasher. Dordrecht: Springer.

Pike, Nelson. 1970. *God and timelessness.* London: Routledge & Kegan Paul.

Pink, Thomas. 2004. *Free will.* Oxford: Oxford University Press.

Plantinga, Alvin. 1982. *The nature of necessity.* Oxford: Clarendon Press.

Plato. 1922. *Platon Staten.* Trans. Claes Lindskog. Nora: Nya Doxa.

Popper, Karl. 2005. *The logic of scientific discovery.* London/New York: Routledge/Taylor and Francis.

Premack, David. 1988. The chimpanzee have a theory of mind? Revisited. In *Machiavellian intelligence: Social expertise and the evolution of intellect in monkeys, apes and humans*, ed. Richard Byrne and Andrew Whiten. Oxford: Oxford University Press.

Proctor, James, D. 2005. Frans de Waal and Evan Thompson, Interviewed by James Proctor. *Journal of Consciousness Studies* 12(7):38–54.

Pryor, Karin, and Ingrid Kang Shallenberger. 1998. Social structure in spotted dolphin (Stenella attenuate) in the Tuna Purse Seine Fishery in the Eastern Tropical Pacific. In *Dolphin societies, discoveries and puzzles*, ed. Karin Pryor and Kenneth S. Norris. Berkeley/Los Angeles: University of California Press.

Puddefoot, John. 1996. *God and the mind machine: Computer, artificial intelligence and the human soul*. London: SPCK.

Putnam, Hilary. 1981. *Reason, truth and history*. Cambridge: Cambridge University Press.

Putnam, Hilary. 1994. *Words and life*. Cambridge: Harvard University Press.

Quine, Willard Van Orman. 1960. *Word and object*. Cambridge: M.I.T. Press.

Raichle, Marcus E. 2010. The brain's dark energy. *Scientific American* 1: 28–33.

Reiss, Diana. 2011. *The dolphins in the mirror. Exploring dolphin minds and saving dolphin lives*. Boston/New York: Houghton Mifflin Harcourt.

Reiss, Diana, and Lori Marino. 2001. Mirror self-recognition in the bottlenose dolphin: A case of cognitive convergence. *PNAS* 98(10): 5937–5942.

Reyes, Hermán. 2007. The worst scars are in the mind: Psychological torture. *International Review* 89(867): 591–616.

Richerson, Peter J., and Robert Boyd. 2006. *Not by genes alone: How culture transformed human evolution*. London: The University of Chicago Press.

Ring, Kenneth. 1985. *Heading toward omega. In search of the meaning of near-death experience*. New York: HarperCollinsPublishers.

Ristau, Carolyn A., ed. 1991. *Cognitive ethology: The minds of other animals*. Volume compiled in honor of Donald R. Griffin. Hillsdale: Lawrence Erlbaum.

Rizzolatti, Giacomo, and Laila Craigtero. 2004. The mirror neuron system. *Annual Review of Neuroscience* 27: 169–192.

Rizzolatti, Giacomo, and Corrado Sinigaglia. 2008. *Mirrors in the brain: How our minds share actions and emotions*. New York: Oxford University Press.

Rizzolatti, Giacomo, Luciano Fadiga, Vittorio Gallese, and Leonardo Fogassi. 1996. Premotor cortex and the recognition of motor actions. *Cognitive Brain Research* 3: 131–141.

Rorty, Richard. 1995. Deconstruction. In *From fromalism to postructuralism, the Cambridge history of literal criticism,* ed. Raman Selden, vol. 8, 166–196.

Rosin, Hanna. 2014. *When men are raped*. Doublex, April 29. http://www.slate.com/articles/double_x/doublex/2014/04/male_rape_in_america_a_new_study_reveals_that_men_are_sexually_assaulted.html. Accessed 5 Feb 2016.

Rottschaefer, William Andrew. 2001. Emerging naturalism: William Rottschaefer and Willem Drees in converstation: How to make naturalism safe for supernaturalism: An evalutation of Willem Drees's supernaturalistic naturalism. *Zygon* 36(3): 407–453.

Rousseau, Jean-Jacques. 2004. *Discourse on the origin of inequality*. Mineola/New York: Dover Publ. Inc.

Rubin, Charles T. 2003. Artificial intelligence and human nature. *The Atlantis. A Journal of Technology & Society* 1: 88–100.

Ruby, Perrine, and Jean Decety. 2004. How would you feel versus how do you think she would feel. A neuroimaging study of perspective-taking with social emotions. *Journal of Cognitive Neuroscience* 16: 988–999.

Runehov, Anne L.C. 2006. A being or to be? Philosophical thoughts about future research on neuroscience and religions and the need for interdisciplinarity. *EJST: European Journal of Science and Theology* 2(1): 55–66.

Runehov, Anne L.C. 2007. *SACRED or NEURAL? The potential of neuroscience to explain religious experience*. Göttingen: Vandenhoeck & Ruprecht GmbH & Co. KG.

Runehov, Anne L.C. 2008a. Neuroscientific research on the self: A case for panentheism? *Metanexus Institute.* http://www.metanexus.net/conference2008/articles/Default.aspx?id=10513

Runehov, Anne L.C. 2008b. Cyborgs and smart mice: How human can they get? *EJST: European Journal of Science and Theology* 4:1, 5–17.

Runehov, Anne L.C. 2009a. The emergent threefold self: A response to Roger Trigg. In *Theology, evolution and the mind*, ed. Neil Spurway, 119–130. Cambridge: Cambridge Scholar Press.

Runehov, Anne L.C. 2009b. God, Richard Dawkins and Carolus Linnaeus: A case for panentheism. In *Linnaeus and homo religiosus: Biological roots of religious awareness and human identity*, ed. Carl-Reinhold Bråkenhielm, 79–91. Uppsala: Acta Universitatis Upsaliensis.

Runehov, Anne L.C. 2010. Why evil will not go away: A philosophical analysis of the complexity of the problem of evil. In *SSTh XII: Studies in science & theology,* eds. Dirk Evers, Antje Jackelén and Taede A. Smedes, 325–344. Tübingen: Forum Scientiarum.

Runehov, Anne L.C. 2012a. The uniqueness of human social ontology. In *Pensamiento: Cienta, Filosofía y Religión.* Madrid: Universidag Pontificia Comillas, seric especial n. 5. vol. 67:254, 708–721.

Runehov, Anne L.C. 2012b. Imago Dei and Simulatio or Imitatio Dei: A philosophical essay on empathy. *Theology and Science* 10(4):411–431. Routledge/Taylor & Francis Group.

Runehov Anne L.C. 2014. God or ultimate reality in theory and practice. In *Neuroscience, consciousness and spirituality,* eds. Stefan Schmidt, and Harald Walach, 297–317. New York/Berlin: Springer.

Runehov Anne L.C. 2016a. How to understand time in relation to timeless divine action in a time-dependent world? In *Embracing the ivory tower and stained glass window,* ed. Jennifer Baldwin, 27–45. Heidenberg/New York/Dordrecht/London: Springer.

Runehov Anne L.C. 2016b. Cognitive or affective? A philosophical analysis of modes of understanding compassion. In *Issues in science and theology: Do emotions shape the world?,* eds. Michael Fuller, Dirk Evers, Anne Runehov and Knut-Willy Saether. Springer, forthcoming.

Runehov, Anne L.C. and Hans Ferdinand Angel. 2014. The process of believing: Revisiting the problem of justifying beliefs. In *SSTh, studies in science and theology*, eds. Dirk Evers, Michael Fuller, Antje Jackelén and Knut-Willy Saether, Tübingen: Forum Scientiarum vol. 14, 205–222.

Russell, Bertrand. 1915. On the experience of time. *The Monist* 25: 212–233.

Sabom, M.B. 1998. *Light and death: One doctor's fascinating account of near-death experiences.* Grand Rapids: Zondervan Publishing House.

Sandbeck, Lars. 2007. *Mig og Evigheden: Johannes Sløks Religionsfilosofi.* København: Forlaget Anis.

Sartre, Jean Paul. 1943. *L'être et le Néant.* Paris: Tel Gallimard.

Schopenhauer, A. 1992. *Världen som vilja och föreställning.* Nora: Nya Doxa.

Schrödinger, Erwin. 1967. *What is life? With mind and matter and autobiographical sketches.* Cambridge/New York: Cambridge University Press.

Searle, John R. 1980. Minds, brains and programs. *Behavior and Brain Sciences* 3: 417–424.

Searle, John R. 1983. *Intentionality: An essay in the philosophy of mind.* Cambridge: Cambridge University Press.

Searle, John R. 1990. Is the brain a computer program? *Scientific American* 261(1): 26–31.

Searle, John R. 2007. *Freedom & neurobiology. Reflecting on free will, language and political power.* New York/Chichester: Columbia University Press.

Searle, John R. 2008. *Philosophy in a New century: Selected essays.* Cambridge: Cambridge University Press.

Searle, John R. 2010. *Making the social world: The structure of human civilization.* Oxford/New York: Oxford University Press.

Seigel, Jerrold. 2005. *The idea of the self: Thought and experience in Western Europe since the seventeenth century.* Cambridge/New York: Cambridge University Press.

Seitz, Rüdiger J., R. Schäfer, D. Scherfeld, S. Friederichs, K. Popp, H.-J. Wittsach, N.P. Azari, and M. Franz. 2008. Valuating other people's emotional face expression: A combined functional magnetic resonance imaging and electroencephalography study. *Neuroscience* 152: 713–722.

Sellars, Roy Wood. 1927. Why naturalism and not materialism. *Philosophical Review* 36(3): 216–225.

Semendeferi, K., A. Lu, N. Schenker, and H. Damasio. 2002. Humans and great apes share a large frontal cortex. *Nature Neuroscience* 5(3): 272–276.

Semendeferi, Katarina, Kate Teffer, Dan P. Buxhoeveden, Min S. Park, Sebastian Bludau, Katrin Amunts, Katie Travis, and Josephy Buckwalter. 2011. Spatial organization of neurons in the frontal pole sets humans apart from great Apes. *Cerebral Cortex* 21:1485–1497.

Seneca, Lucius Annasus. 1928–1935. *Seneca's essays*. Trans. John W. Basore, The Loeb Classical Library: London: W. Heinemann, (3 vols), vol. I. Available on http://www.stoics.com/seneca_essays_book_1.html. Accessed 5 Feb 2016.

Shamay-Tsoory, Simon G., Rachel Tomer, B.D. Berger, D. Goldsher and J. Aharon-Peretz. 2005. The neural correlates of understanding the other's distress: A positron emission tomography investigation of accurate empathy. *NeuroImage* 27(2);468–472.

Shaw, P., E.J. Lawrence, C. Radbourne, J. Bramham, C.E. Polkey, and A.S. David. 2004. The impact of early and late damage to the human amygdala on 'theory of mind' reasoning. *Brain* 127: 1535–1548.

Siegenthaler, Hermann. 1980. *Die Entwicklung des Gottesbildes bei Kindern und Jugendlichen.* Württenberg: Fachgemeinschaft evangelishcer Religionslehrer, Baden e.: Fachverbund evangelischen Religionslehrer.

Sitaram, Ranganatha. 2013. *Neuroimaging meditation*. http://www.bme.ufl.edu/labs/wp-content/blogs.dir/12/files/2013/10/NeuroimagingMeditation.pdf. Accessed 1 Feb 2016.

Skirry, Justin. 2016. René Descartes, the mind-body distinction. *Internet encyclopedia of philosophy and its authors*. http://www.iep.utm.edu/descmind/. Accessed 26 Jan 2016.

Smith, Adam. [1759] 2005. *The theory of moral sentiments*. London: A. Miller. Made available through Sálvio Marcelo Soares, https://www.ibiblio.org/ml/libri/s/SmithA_MoralSentiments_p.pdf. Accessed 5 Feb 2016.

Smith, J.D., J. Schull, J. Strote, K. McGee, R. Egnor, and L. Erb. 1995. The uncertain response in the bottlenosed dolphin (Tursiops truncatus). *Journal of Experimental Psychology. General* 124: 391–408.

Spinoza, Baruch. 1983. *Etik*. Göteborg: Daidalos.

Sprengelmeyer, R., A.W. Young, U. Schroeder, P.G. Grossenbacher, J. Federlein and T. Buttner. 1999. Knowing no fear. *Proceedings of the Royal Society* (Series B: Biology), 266:2451–2456.

Spurway, Neil. 2009. *Theology, evolution and the mind*. Newcastle upon Tyne: Cambridge Scholar Publishing.

Stein, Edith. 1989. *On the problem of empathy*. Washington, DC: ICS publications.

Stemple, Lara, and Ilan H. Meyer. 2014. The sexual victimization of men in America: New data challenge old assumptions. *American Journal of Public Health* 104(6): 19–26.

Stenmark, Mikael. 2015. Competing conception of god: The personal god versus the god beyond being. *Religious Studies* 51(2): 205–220.

Stump, Eleonora. 2003. *Aquinas*. London: Routledge.

Stump, Eleonore, and Norman Kretzmann. 1981. Eternity. *The Journal of Philosophy* 78(8): 429–458.

Stump, Eleonore, and Norman Kretzmann. 1987. Atemporal duration, a reply to Fitzgerald. *The Journal of Philosophy* 84(4): 214–219.

Stuss, D.T., G.G. Jr. Gallup, and M.P. Alexander. 2001. The frontal lobes are necessary for 'theory of mind'. *Brain* 124(2):279–286.

Tart, Charles T. 2006. Current status of transpersonal psychology. *Journal of Consciousness Studies* 13(4): 83–87.

Taylor Sloan, J. 2014. *Science shows how piano players' brains are actually different from everybody elses.* http://mic.com/articles/91329/science-shows-how-piano-players-brains-are-actually-different-from-everybody-elses#.fPcL3XMDR. Accessed 1 Feb 2016.

The Holy Bible. New International Version. London/Sydney/Auckland/Toronto: Hodder and Stoughton.

Thompson, Evan. 2001. Empathy and consciousness. *Journal of Conscious Studies* 8(5–7): 1–32.

Till, J.E., and E.A. McCulloch. 1961. A direct measurement of the radiation sensitivity of normal mouse bone marrow cells. *Radiation Research* 14: 213–222.

Tomassello, Michael, M. Carpenter, J. Call, T. Behne, and H. Moll. 2005. Understanding and sharing intentions: The origins of cultural cognition. *Behavioral and Brain Sciences* 28: 675–735.

Toom, Tarmo. 2007. *Classical trinitarian theology*. New York/London: t & t clark.

Travis, Frederick, and Craig Pearson. 2000. Pure consciousness: Distinct phenomenological and physiological correlates of 'consciousness' itself. *International Journal of Neuroscience* 100(1–4): 77–89.

Trigg, Roger. 2009. Are we ghost or machines? In *Theology, evolution and the mind*, ed. Neil Spurway, 104–118. Cambridge: Cambridge Scholar Publishing.

Tummolini, Luca, Christiano Catelfranchi, and Hannes Rakoczy. 2006. Pretend play and the develpment of collective intentionality. *Cognitive Systems Research* 7: 113–127.

Tuomela, Raimo. 2013. *Social ontology, collective intentionality and group agents*. Oxford/New York: Oxford University Press.

Turing, Alen M. 1990. Computing machinery and intelligence. In *The philosophy of artificial intelligence*, ed. Margareth Boden, 40–66. Oxford: Oxford University Press.

Tye, Michael. 1999. Phenomenal consciousness: The explanatory gap as a cognitive illusion. *Mind* 108: 705–725.

Underhill, Evelyn. 1912. *Mysticism, a study in the nature and development of man's spiritual consciousness*. New York: Forgotten Books.

Van Huyssteen, Wentzel. 2006. *Alone in the world? Human uniqueness in science and theology*, The Gifford lectures. Cambridge: Eerdmans Publishing Göttingen/Vandenhoeck & Ruprecht.

Van Lommel, P., R. van Wees, V. Meywers, and I. Eifferich. 2001. Near-death experience in survivors of cardia arrest: A prospective study in the Netherlands. *The Lancet* 358: 2039–2045.

Walach, Harald. 2007. Mind – Body – Spirituality. *Mind & Matter* 5(2): 215–240.

Walach, Harald, and Anne L.C. Runehov. 2010. The epistemology status of transpersonal psychology – The data-base argument revisited. *Journal of Consciousness Studies* 17(1–2): 145–165.

Ward, A. 2005. Defending ethical naturalism: The roles of cognitive science and pragmatism. *Zygon* 40(1): 201–220.

Watts, Pauline Moffitt. 1983. *Nicolaus Cusanus: A fifteenth-century vision of man*. Leiden: E.J. Brill.

Wheeler, William Morton. 1928. *Emergent evolution and the development of societies*. New York: W.W. Morton & Company Inc.

Whiten, Andrew. 2000. Chimpanzee cognition and the question of mental re-representation. In *Metarepresentations: A multidisciplinary perspective*, ed. Sperber Dan. Oxford: Oxford University Press.

Whiten, Andrew, and Richard W. Byrne (eds.). 1997. *Machiavellian intelligence II, extensions and evaluations*. Cambridge/New York: Cambridge University Press.

Wigner, Eugene P. 1967. *Symmetries and reflections*. Bloomington: Indiana University Press.

Wiles, Maurice F. 2012. *God's action in the world: The Bampton lectures for 1986*. London: SCM Press.

Winograd, Terry. 1991. Thinking machines: Can there be? Are we? In *The boundaries of humanity: Humans, animals, machines*, ed. J. Sheehan and M. Sosna, 198–223. Berkeley: University of California Press.

Wiser, James L. 1983. *Political philosophy: A history of the search for order*. Englewood Cliffs: Prentice-Hall Inc.

Wittgenstein, Ludwig. 1953. Philosophical investigations. Trans. G.E.M. Anscombe. Oxford: Basil Blackwell.
Wittgenstein, Ludwig. 1992. *Tractatus logico-philosophicus*. Stockholm: Thales.
Xinzhong, Yao. 2000. *An introduction to confucianism*. New York: Canbridge University Press.
Yandell, Keith. 1993. *The epistemology of religious experience*. Cambridge: Cambridge University Press.
Zagzebski, Linda Trinkaus. 2004. *Divine motivation theory*. Cambridge: Cambridge University Press.
Zangwill, Nick. 2006. Explaining human cruelty. *Behavior and Brain Sciences* 29: 245–246.
Zukav, Gary. 1979. *De Dansande Wu Li-mästarna*. Stockholm: Askild & Kärnekull.

Webpages

Androids. see, http://roboticsandtechnology.blogspot.se/p/humanoid-robot.html. Accessed 30 Sep 2015.
Bohr, Niels, http://www.newworldencyclopedia.org/entry/Niels_Bohr
Einstein, Albert. http://www.goodreads.com/quotes/tag/compassion. Accessed 5 Feb 2016.
Exact what is time? Ancient Philosophy. http://www.exactlywhatistime.com/ancient-philosophy. Accessed 8 Feb 2016.
Newton, Isaac. 2004. http://plato.stanford.edu/entries/newton-stm/. Accessed 8 Feb 2016.

MIX
Papier aus verantwortungsvollen Quellen
Paper from responsible sources
FSC® C105338

If you have any concerns about our products,
you can contact us on
ProductSafety@springernature.com

In case Publisher is established outside the EU,
the EU authorized representative is:
Springer Nature Customer Service Center GmbH
Europaplatz 3, 69115 Heidelberg, Germany

Printed by Libri Plureos GmbH
in Hamburg, Germany